W9-AYM-801

The Intelligence Trap

The Intelligence Trap

Why Smart People Make Dumb Mistakes

DAVID ROBSON

W. W. NORTON & COMPANY
Independent Publishers Since 1923
NEW YORK • LONDON

First American Edition 2019

First published in Great Britain under the title *The Intelligence Trap: Why Smart People Do Stupid Things and How to Make Wiser Decisions*

Printed in the United States of America

Manufacturing by Lake Book
Production manager: Anna Oler

ISBN 978-0-393-65142-3

W. W. Norton & Company, Inc., 500 Fifth Avenue, New York, N.Y. 10110
www.wwnorton.com

W. W. Norton & Company Ltd., 15 Carlisle Street, London W1D 3BS

1 2 3 4 5 6 7 8 9 0

To my parents, and to Robert

Contents

Part 4—The folly and wisdom of the crowd: How teams and organizations can avoid the intelligence trap

Introduction

Venture into the darker recesses of the internet, and you may come across the views of a man named Kary. If he is to be believed, he has some unique insights that could change the world order.[1]

He suspects he was abducted by an alien near the Navarro River, California, for instance, after encountering a strange being who took the form of a glowing raccoon with "shifty black eyes." He can't actually remember what happened "after the little bastard" gave him a "courteous greeting"; the rest of the night is a complete blank. But he strongly suspects it involved extra-terrestrial life. "There are a lot of mysteries in the valley," he writes, cryptically.

He's also a devoted follower of astrology. "Most [scientists] are under the false impression that it is non-scientific and not a fit subject for their serious study," he huffs in a long rant. "They are dead wrong." He thinks it's the key to better mental health treatment and everyone who disagrees has "their heads firmly inserted in their asses." Besides these beliefs in ET and star signs, Kary also thinks that people can travel through the ether on the astral plane.

Things take a darker turn when Kary starts talking about politics. "Some of the big truths voters have accepted have little or no scientific basis," he claims. This includes "the belief that AIDS is caused by HIV virus" and "the belief that the release of CFCs into the atmosphere has created a hole in the ozone layer."

Needless to say, these ideas are almost universally accepted by scientists—but Kary tells his readers that they are just out for money. "Turn off your TV. Read your elementary science textbooks," he implores. "You need to know what they are up to."

I hope I don't have to tell you that Kary is wrong.

The web is full of people with groundless opinions, of course—but we don't expect astrologers and AIDS denialists to represent the pinnacle of intellectual achievement.

Yet Kary's full name is Kary Mullis, and far from being your stereotypically ill-informed conspiracy theorist, he is a Nobel Prize-winning scientist—placing him alongside the likes of Marie Curie, Albert Einstein and Francis Crick.

Mullis was awarded the prize for his invention of the polymerase chain reaction—a tool that allows scientists to clone DNA in large quantities. The idea apparently came to him during a flash of inspiration on the road in Mendocino County, California, and many of the greatest achievements of the last few decades—including the Human Genome Project—hinged on that one moment of pure brilliance. The discovery is so important that some scientists even divide biological research into two eras—before and after Mullis.

There can be little doubt that Mullis, who holds a PhD from the University of California, Berkeley, is incredibly intelligent; his invention can have only come from a lifetime dedicated to understanding the extraordinarily complex processes inside our cells.

But could the same genius that allowed Mullis to make that astonishing discovery also explain his beliefs in aliens and his AIDS denialism? Could his great intellect have also made him incredibly stupid?

~

This book is about why intelligent people act stupidly—and why in some cases they are even more prone to error than the average person. It is also about the strategies that we can all employ to avoid the same mistakes: lessons that will help anyone to think more wisely and rationally in this post-truth world.

You don't need to be a Nobel Prize winner for this to apply to you. Although we will discover the stories of people like Mullis, and Paul Frampton, a brilliant physicist who was fooled into carrying two kilograms of cocaine across the Argentinian border, and Arthur Conan Doyle, the famed author who fell for two teenagers' scams, we will also see how the same flaws in thinking can lead anyone of more than average intelligence astray.

Like most people, I once believed that intelligence was synonymous with good thinking. Since the beginning of the twentieth century, psychologists have measured a relatively small range of abstract skills—factual recall, analogical reasoning, and vocabulary—in the belief that they reflect an innate general intelligence that underlies all kinds of learning, creativity, problem solving, and decision making. Education is then meant to build on that "raw" brainpower, furnishing us with more specialized knowledge in the arts, the humanities, and the sciences that will also be crucial for many professions. The smarter you are—according to these criteria—the more astute your judgment.

But as I began working as a science journalist, specializing in psychology and neuroscience, I noticed the latest research was revealing some serious problems with these assumptions. Not only do general intelligence and academic education fail to protect us from various cognitive errors; smart people may be even *more* vulnerable to certain kinds of foolish thinking.

Intelligent and educated people are less likely to learn from their mistakes, for instance, or take advice from others. And when they do err, they are better able to build elaborate arguments to justify their reasoning, meaning that they become more and more dogmatic in their views. Worse still, they appear to have a bigger "bias blind spot," meaning they are less able to recognize the holes in their logic.

Intrigued by these results, I began looking further afield. Management scientists, for example, have charted the ways that poor corporate cultures—aimed to increase productivity—can amplify irrational decision making in sports teams, businesses, and government organizations. As a result, you can have whole teams built of incredibly intelligent people, who nevertheless make incredibly stupid decisions.

The consequences are serious. For the individual, these errors can influence our health, well-being, and professional success. In our courts it is leading to serious miscarriages of justice. In hospitals, it may be the reason that 15 percent of all diagnoses are wrong, with more people dying from these mistakes than diseases like breast cancer. In business, it leads to bankruptcy and ruin.[2]

The vast majority of these mistakes cannot be explained by a lack of knowledge or experience; instead, they appear to arise from the particular, flawed mental habits that come with greater intelligence, education, and professional expertise. Similar errors can lead spaceships to crash, stock markets to implode, and world leaders to ignore global threats like climate change.

Although they may seem to be unconnected, I found that some common processes underlie all these phenomena: a pattern that I will refer to as the intelligence trap.[3]*

Perhaps the best analogy is a car. A faster engine *can* get you places more quickly if you know how to use it correctly. But simply having more horsepower won't guarantee that you will arrive at your destination safely. Without the right knowledge and equipment—the brakes, the steering wheel, the speedometer, a compass, and a good map—a fast engine may just lead to you driving in circles—or straight into oncoming traffic. And the faster the engine, the more dangerous you are.

In exactly the same way, intelligence can help you to learn and recall facts, and process complex information quickly, but you also need the necessary checks and balances to apply that brainpower correctly. Without them, greater intelligence can actually make you *more* biased in your thinking.

Fortunately, besides outlining the intelligence trap, recent psychological research has also started to identify those additional mental qualities that can keep us on track. As one example, consider the following deceptively trivial question:

> *Jack is looking at Anne but Anne is looking at George. Jack is married but George is not. Is a married person looking at an unmarried person?*

---

* It's worth noting that Edward de Bono first used the term "intelligence trap" to describe some of the potential disadvantages of greater general intelligence in his books on lateral thinking and creativity. Similarly, the Harvard University psychologist David Perkins refers to "intelligence traps" in his book *Outsmarting IQ* (Simon & Schuster, 1995). Perkins's ideas, in particular, have informed some elements of my argument, and I would thoroughly recommend reading his work.

Yes, No, or Cannot Be Determined?

The correct answer is "yes"—but the vast majority of people say "cannot be determined."

Don't feel disheartened if you didn't initially get it. Many Ivy League students get it wrong, and when I published this test in *New Scientist* magazine, we had an unprecedented number of letters claiming that the answer was a mistake. (If you still can't see the logic, I'd suggest drawing a diagram, or see p. 270.)

The test measures a characteristic known as cognitive reflection, which is the tendency to question our own assumptions and intuitions, and people who score badly on this test are more susceptible to bogus conspiracy theories, misinformation and fake news. (We'll explore this some more in Chapter 6.)

Besides cognitive reflection, other important characteristics that can protect us from the intelligence trap include intellectual humility, actively open-minded thinking, curiosity, refined emotional awareness, and a growth mind-set. Together, they keep our minds on track and prevent our thinking from veering off a proverbial cliff.

This research has even led to the birth of a new discipline—the study of "evidence-based wisdom." Once viewed with skepticism by other scientists, this field has blossomed in recent years, with new tests of reasoning that better predict real-life decision-making than traditional measures of general intelligence. We are now even witnessing the foundation of new institutions to promote this research—such as the Center for Practical Wisdom at the University of Chicago, which opened in June 2016.

Although none of these qualities are measured on standard academic tests, you don't need to sacrifice any of the benefits of having high general intelligence to cultivate these other thinking styles and reasoning strategies; they simply help you to apply your intelligence more wisely. And unlike intelligence, they can be trained. Whatever your IQ, you can learn to think more wisely.

~

This cutting-edge science has a strong philosophical pedigree. An early discussion of the intelligence trap can even be found at Socrates's trial in 399 BC.

According to Plato's account, Socrates's accusers claimed that he had been corrupting Athenian youth with evil "impious" ideas. Socrates denied the charges, and instead explained the origins of his reputation for wisdom—and the jealousy behind the accusations.

It started, he said, when the Oracle of Delphi declared that there was no one in Athens who was wiser than Socrates. "What can the god be saying? It's a riddle: what can it mean?" Socrates asked himself. "I've no knowledge of my being wise in any respect, great or small."

Socrates's solution was to wander the city, seeking out the most respected politicians, poets, and artisans to prove the oracle wrong—but each time, he was disappointed. "Because they were accomplished in practising their skill, each one of them claimed to be wisest about other things too: the most important ones at that—and this error of theirs seemed to me to obscure the wisdom they did possess. . . .

"Those with the greatest reputations," he added, "seemed to me practically the most deficient, while others who were supposedly inferior seemed better endowed when it came to good sense."

His conclusion is something of a paradox: he is wise precisely because he recognized the limits of his own knowledge. The jury found him guilty nonetheless, and he was sentenced to death.[3]

The parallels with the recent scientific research are striking. Replace Socrates's politicians, poets, and artisans with today's engineers, bankers, and doctors, and his trial almost perfectly captures the blind spots that psychologists are now discovering. (And like Socrates's accusers, many modern experts do not like their flaws being exposed.)

But as prescient as they are, Socrates's descriptions don't quite do the new findings justice. After all, none of the researchers would deny that intelligence and education are essential for good thinking. The problem is that we often don't use that brainpower correctly.

For this reason, it is René Descartes who comes closest to the modern understanding of the intelligence trap. "It is not enough to possess a good mind; the most important thing is to apply it correctly," he wrote in his *Discourse on the Method* in 1637. "The greatest minds are capable of the greatest vices as well as the greatest virtues; those who go forward but very slowly can get further, if they always follow the right road, than those who are in too much of a hurry and stray off it."[4]

The latest science allows us to move far beyond these philosophical musings, with well-designed experiments demonstrating the precise reasons that intelligence can be a blessing and a curse, and the specific ways to avoid those traps.

~

Before we begin this journey, let me offer a disclaimer: there is much excellent scientific research on the theme of intelligence that doesn't find a place here. Angela Duckworth at the University of Pennsylvania, for instance, has completed groundbreaking work on the concept of "grit," which she defines as our "perseverance and passion for long-term goals," and she has repeatedly shown that her measures of grit can often predict achievement better than IQ. It's a hugely important theory, but it's not clear that it could solve the particular biases that appear to be exaggerated with intelligence; nor does it fall under the more general umbrella of evidence-based wisdom that guides much of my argument.

When writing *The Intelligence Trap*, I've restricted myself to three particular questions. Why do smart people act stupidly? What skills and dispositions are they missing that can explain these mistakes? And how can we cultivate those qualities to protect us from those errors? And I have examined them at every level of society, starting with the individual and ending with the errors plaguing huge organizations.

Part 1 defines the problem. It explores the flaws in our understanding of intelligence and the ways that even the brightest minds can backfire—from Arthur Conan Doyle's dogged beliefs in fairies to the FBI's flawed investigation into the Madrid bombings of 2004—and the reasons that knowledge and expertise only exaggerate those errors.

Part 2 presents solutions to these problems by introducing the new discipline of "evidence-based wisdom," which outlines those other thinking dispositions and cognitive abilities that are crucial for good reasoning, while also offering some practical techniques to cultivate them. Along the way, we will discover why our intuitions often fail and the ways we can correct those errors to fine-tune our instincts. We will also explore strategies to avoid misinformation and fake news, so that we can be sure that our choices are based on solid evidence rather than wishful thinking.

Part 3 turns to the science of learning and memory. Despite their brainpower, intelligent people sometimes struggle to learn well, reaching a kind of plateau in their abilities that fails to reflect their potential. Evidence-based wisdom can help to break that vicious cycle, offering three rules for deep learning. Besides helping us to meet our own personal goals, this cutting-edge research also explains why East Asian education systems are already so successful at applying these principles, and the lessons that Western schooling can learn from them to produce better learners and wiser thinkers.

Finally, Part 4 expands our focus beyond the individual, to explore the reasons that talented groups act stupidly—from the failings of the England football team to the crises of huge organizations like BP, Nokia, and NASA.

The great nineteenth-century psychologist William James reportedly said that "a great many people think they are thinking when they are merely rearranging their prejudices." *The Intelligence Trap* is written for anyone, like me, who wants to escape that mistake—a user's guide to both the science, and art, of wisdom.

PART 1

The downsides of intelligence: *How a high IQ, education, and expertise can fuel stupidity*

1

The rise and fall of the Termites: *What intelligence is—and what it is not*

As they nervously sat down for their tests, the children in Lewis Terman's study can't have imagined that their results would forever change their lives—or world history.* Yet each, in their own way, would come to be defined by their answers, for good and bad, and their own trajectories would permanently change the way we understand the human mind.

One of the brightest was Sara Ann, a six-year-old with a gap between her front teeth, and thick spectacles. When she had finished scrawling her answers, she casually left a gumdrop in between her papers—a small bribe, perhaps, for the examiner. She giggled when the scientist asked her whether "the fairies" had dropped it there. "A little girl gave me two," she explained sweetly. "But I believe two would be bad for my digestion because I am just well from the flu now." She had an IQ of 192—at the very top of the spectrum.[1]

Joining her in the intellectual stratosphere was Beatrice, a precocious little girl who began walking and talking at seven months. She had read 1,400 books by the age of ten, and her own poems were apparently so mature that a local San Francisco newspaper claimed they had "completely fooled an English class at Stanford," who mistook them for the works of Tennyson. Like Sara Ann, her IQ was 192.[2]

* The stories of the following four children are told in much greater detail, along with the lives of the other "Termites," in Shurkin, J. (1992), *Terman's Kids: The Groundbreaking Study of How the Gifted Grow Up*, Boston, MA: Little, Brown.

Then there was eight-year-old Shelley Smith—"a winsome child, loved by everyone"; her face apparently glowed with suppressed fun.[3] And Jess Oppenheimer—"a conceited, egocentric boy" who struggled to communicate with others and lacked any sense of humor.[4] Their IQs hovered around 140—just enough to make it into Terman's set, but still far above average, and they were surely destined for great things.

Up to that point, the IQ test—still a relatively new invention—had been used mostly to identify people with learning difficulties. But Terman strongly believed that these few abstract and academic traits—such as memory for facts, vocabulary, and spatial reasoning skills—represent an innate "general intelligence" that underlies all your thinking abilities. Irrespective of your background or education, this largely innate trait represented a raw brainpower that would determine how easily you learn, understand complex concepts, and solve problems.

"There is nothing about an individual as important as his IQ," he declared at the time.[5] "It is of the highest 25% of our population, and more especially to the top 5%, that we must look for the production of leaders who will advance science, art, government, education, and social welfare generally."

By tracking the course of their lives over the subsequent decades, he hoped that Sara Ann, Beatrice, Jess, and Shelley and the other "Termites" were going to prove his point, predicting their success at school and university, their careers and income, and their health and well-being; he even believed that IQ would predict their moral character.

The results of Terman's studies would permanently establish the use of standardized testing across the world. And although many schools do not explicitly use Terman's exam to screen children today, much of our education still revolves around the cultivation of that narrow band of skills represented in his original test.

If we are to explain why smart people act foolishly, we must first understand how we came to define intelligence in this way, the abilities this definition captures, and some crucial aspects of thinking that it misses—skills that are equally essential for creativity and pragmatic problem solving, but which have been completely neglected in our education system. Only then can we begin to contemplate the origins of the intelligence trap—and the ways it might also be solved.

We shall see that many of these blind spots were apparent to contemporary researchers as Terman set about his tests, and they would become even more evident in the triumphs and failures of Beatrice, Shelley, Jess, Sara Ann, and the many other "Termites," as their lives unfolded in sometimes dramatically unexpected ways. But thanks to the endurance of IQ, we are only just getting to grips with what this means and the implications for our decision making.

Indeed, the story of Terman's own life reveals how a great intellect could backfire catastrophically, thanks to arrogance, prejudice—and love.

~

As with many great (if misguided) ideas, the germs of this understanding of intelligence emerged in the scientist's childhood.

Terman grew up in rural Indiana in the early 1880s. Attending a "little red schoolhouse," a single room with no books, the quiet, red-headed boy would sit and quietly observe his fellow pupils. Those who earned his scorn included a "backward" albino child who would only play with his sister, and a "feeble-minded" eighteen-year-old still struggling to grasp the alphabet. Another playmate, "an imaginative liar," would go on to become an infamous serial killer, Terman later claimed—though he never said which one.[6]

Terman, however, knew he was different from the incurious children around him. He had been able to read before he entered that bookless schoolroom, and within the first term the teacher had allowed him to skip ahead and study third-grade lessons. His intellectual superiority was only confirmed when a traveling salesman visited the family farm. Finding a somewhat bookish household, he decided to pitch a volume on phrenology. To demonstrate the theories it contained, he sat with the Terman children around the fireside and began examining their scalps. The shape of the bone underneath, he explained, could reveal their virtues and vices. Something about the lumps and bumps beneath young Lewis's thick ginger locks seemed to have particularly impressed him. This boy, he predicted, would achieve "great things."

"I think the prediction probably added a little to my self-confidence and caused me to strive for a more ambitious goal than I might otherwise have set," Terman later wrote.[7]

By the time he was accepted for a prestigious position at Stanford University in 1910, Terman would long have known that phrenology was a pseudoscience; there was nothing in the lumps of his skull that could reflect his abilities. But he still had the strong suspicion that intelligence was some kind of innate characteristic that would mark out your path in life, and he had now found a new yardstick to measure the difference between the "feeble-minded" and the "gifted."

The object of Terman's fascination was a test developed by Alfred Binet, a celebrated psychologist in *fin de siècle* Paris. In line with the French Republic's principle of *égalité* among all citizens, the government had recently introduced compulsory education for all children between the ages of six and thirteen. Some children simply failed to respond to the opportunity, however, and the Ministry of Public Instruction faced a dilemma. Should these "imbeciles" be educated separately within the school? Or should they be moved to asylums? Together with Théodore Simon, Binet invented a test that would help teachers to measure a child's progress and adjust their education accordingly.[8]

To a modern reader, some of the questions may seem rather absurd. As one test of vocabulary, Binet asked children to examine drawings of women's faces and judge which was "prettier" (see image below). But many of the tasks certainly did reflect crucial skills that would be essential for their success in later life. Binet would recite a string of numbers or words, for example, and the child had to recall them in the correct order to test their short-term memory. Another question would ask them to form a sentence with three given words—a test of their verbal prowess.

Binet himself was under no illusions that his test captured the full breadth of “intelligence”; he believed our “mental worth” was simply too amorphous to be measured on a single scale and he balked at the idea that a low score should come to define a child’s future opportunities, believing that it could be malleable across the lifetime.[9] “We must protest and react against this brutal pessimism,” he wrote; “we must try to show that it is founded on nothing.”[10]

But other psychologists, including Terman, were already embracing the concept of “general intelligence”—the idea that there is some kind of mental “energy” serving the brain, which can explain your performance in all kinds of problem solving and academic learning.[11] If you are quicker at mental arithmetic, for instance, you are also more likely to be able to read well and to remember facts better. Terman believed that the IQ test would capture that raw brainpower, predetermined by our heredity, and that it could then predict your overall achievement in many different tasks throughout life.[12]

And so he set about revising an English-language version of Binet's test, adding questions and expanding the exam for older children and adults, with questions such as:

If 2 pencils cost 5 cents, how many pencils can you buy for 50 cents?

And:

What is the difference between laziness and idleness?

Besides revising the questions, Terman also changed the way the result was expressed, using a simple formula that is still used today. Given that older children would do better than younger children, Terman first found the average score for each age. From these tables, you could assess a child's "mental age," which, when divided by their actual age and multiplied by 100, revealed their "intelligence quotient." A ten-year-old thinking like a fifteen-year-old would have an IQ of 150; a ten-year-old thinking like a nine-year-old, in contrast, would have an IQ of 90. At all ages, the average would be 100.*

Many of Terman's motives were noble: he wanted to offer an empirical foundation to the educational system so that teaching could be tailored to a child's ability. But even at the test's conception, there was an unsavoury streak in Terman's thinking, as he envisaged a kind of social engineering based on the scores. Having profiled a small group of "hoboes," for instance, he believed the IQ test could be used to separate delinquents from society, before they had even

* For adults, who, at least according to the theory of general intelligence, have stopped developing intellectually, IQ is calculated slightly differently. Your score reflects not your "mental age" but your position on the famous "bell curve." An IQ of 145, for instance, suggests you are in the top 2 percent of the population.

committed a crime.[13] "Morality," he wrote, "cannot flower and fruit if intelligence remains infantile."[14]

Thankfully Terman never realized these plans, but his research caught the attention of the US Army during the First World War, and they used his tests to assess 1.75 million soldiers. The brightest were sent straight to officer training, while the weakest were dismissed from the army or consigned to a labor battalion. Many observers believed that the strategy greatly improved the recruitment process.

Carried by the wind of this success, Terman set about the project that would dominate the rest of his life: a vast survey of California's most gifted pupils. Beginning in 1920, his team set about identifying the *crème de la crème* of California's biggest cities. Teachers were encouraged to put forward their brightest pupils, and Terman's assistants would then test their IQs, selecting only those children whose scores surpassed 140 (though they later lowered the threshold to 135). Assuming that intelligence was inherited, his team also tested these children's siblings, allowing them to quickly establish a large cohort of more than a thousand gifted children in total—including Jess, Shelley, Beatrice, and Sara Ann.

Over the next few decades, Terman's team continued to follow the progress of these children, who affectionately referred to themselves as the "Termites," and their stories would come to define the way we judge genius for almost a century. Termites who stood out include the nuclear physicist Norris Bradbury; Douglas McGlashan Kelley, who served as a prison psychiatrist in the Nuremberg trials; and the playwright Lilith James. By 1959, more than thirty had made it into *Who's Who in America*, and nearly eighty were listed in *American Men of Science*.[15]

Not all the Termites achieved great academic success, but many shone in their respective careers nonetheless. Consider Shelley Smith—"the winsome child, loved by everyone." After dropping out of Stanford University, she forged a career as a researcher and reporter at *Life* magazine, where she met and married the photographer Carl Mydans.[16] Together they traveled around Europe and Asia reporting on political tensions in the build-up to the Second World War; she would later recall days running through foreign streets in a kind of reverie at the sights and sounds she was able to capture.[17]

Jess Oppenheimer, meanwhile—the "conceited, egocentric child" with "no sense of humor"—eventually became a writer for Fred Astaire's radio show.[18] Soon he was earning such vast sums that he found it hard not to giggle when he mentioned his own salary.[19] His luck would only improve when he met the comedian Lucille Ball, and together they produced the hit TV show *I Love Lucy*. In between the script writing, he tinkered with the technology of filmmaking, filing a patent for the teleprompter still used by news anchors today.

Those triumphs certainly bolster the idea of general intelligence; Terman's tests may have only examined academic abilities, but they did indeed seem to reflect a kind of "raw" underlying brainpower that helped these children to learn new ideas, solve problems, and think creatively, allowing them to live fulfilled and successful lives regardless of the path they chose.

And Terman's studies soon convinced other educators. In 1930, he had argued that "mental testing will develop to a lusty maturity within the next half century . . . within a few score years schoolchildren from the kindergarten to the university will be subjected to several times as many hours of testing as would now be thought reasonable."[20] He was right, and many new iterations of his test would follow in the subsequent decades.

Besides examining vocabulary and numerical reasoning, the later tests also included more sophisticated nonverbal conundrums, such as the quadrant on the following page.

The answer relies on your being able to think abstractly and see the common rule underlying the progression of shapes—which is surely reflective of some kind of advanced processing ability. Again, according to the idea of general intelligence, these kinds of abstract reasoning skills are meant to represent a kind of "raw brainpower"—irrespective of your specific education—that underlies all our thinking.

Our education may teach us specialized knowledge in many different disciplines, but each subject ultimately relies on those more basic skills in abstract thinking.

What pattern completes this quadrant?

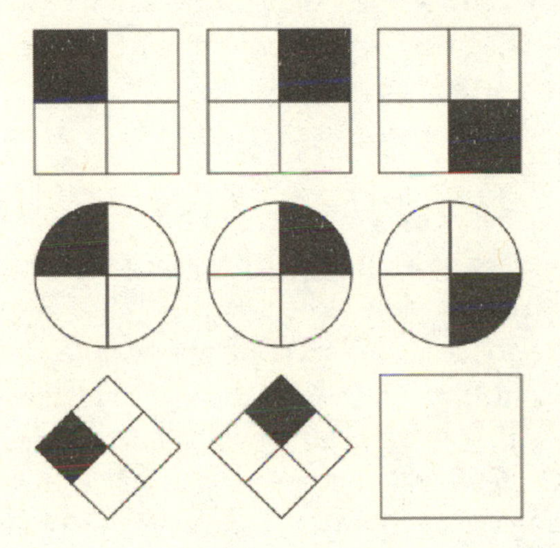

At the height of its popularity, most pupils in the US and the UK were sorted according to IQ. Today, the use of the test to screen young schoolchildren has fallen out of fashion, but its influence can still be felt throughout education and the workplace.

In the US, for instance, the Scholastic Aptitude Test (SAT) used for college admissions was directly inspired by Terman's work in the 1920s. The style of questioning may be different today, but such tests still capture the same basic abilities to remember facts, follow abstract rules, build a large vocabulary, and spot patterns, leading some psychologists to describe them as IQ tests by proxy.

The same is true for many school and university entrance exams and employee recruitment tests—such as Graduate Record Examinations (GREs) and the Wonderlic Personnel Test used for selecting candidates in the workplace. It is a sign of Terman's huge influence that even quarterbacks in the US National Football League take the Wonderlic test during recruitment, based on the theory that greater intelligence will improve the players' strategic abilities on the field.

This is not just a Western phenomenon.[21] Standardized tests, inspired by IQ, can be found in every corner of the globe and in some countries—most notably India, South Korea, Hong Kong, Singapore, and Taiwan—a whole industry of "cram schools" has grown up to coach students for exams like the GRE that are necessary to enter the most prestigious universities.[22] (To give an idea of their importance, in India alone these cram schools are worth $6.4 billion annually.)

Just as important as the exams themselves, however, is the lingering influence of these theories on our attitudes. Even if you are skeptical of the IQ test, many people still believe that those abstract reasoning skills, so crucial for academic success, represent an underlying intelligence that automatically translates to better judgment and decision making across life—at work, at home, in finance, or in politics. We assume that greater intelligence means that you are automatically better equipped to evaluate factual evidence before coming to a conclusion, for instance; it's the reason we find the bizarre conspiracy theories of someone like Kary Mullis to be worthy of comment.

When we do pay lip service to other kinds of decision making that are not measured in intelligence tests, we tend to use fuzzy concepts like "life skills" that are impossible to measure precisely, and we assume that they mostly come to us through osmosis, without deliberate training. Most of us certainly haven't devoted the same time and effort to develop them as we did to abstract thinking and reasoning in education.

Since most academic tests are timed and require quick thinking, we have also been taught that the speed of our reasoning marks the quality of our minds; hesitation and indecision are undesirable, and any cognitive difficulty is a sign of our own failings. By and large, we respect people who think and act quickly, and to be "slow" is simply a synonym for being stupid.

As we shall see in the following chapters, these are all misconceptions, and correcting them will be essential if we are to find ways out of the intelligence trap.

~

Before we examine the limits of the theory of general intelligence, and the thinking styles and abilities that it fails to capture, let's be clear: most psychologists agree that these measures—be they IQ, SATs, GREs, or Wonderlic scores—do reflect something very important about the mind's ability to learn and process complex information.

Unsurprisingly, given that they were developed precisely for this reason, these scores are best at predicting how well you do at school and university, but they are also modestly successful at predicting your career path after education. The capacity to juggle complex information will mean that you find complex mathematical or scientific concepts easier to understand and remember; that capacity to understand and remember difficult concepts might also help you to build a stronger argument in a history essay.

Particularly if you want to enter fields such as law, medicine, or computer programming that will demand advanced learning and abstract reasoning, greater general intelligence is undoubtedly an advantage. Perhaps because of the socioeconomic success that comes with a white-collar career, people who score higher on intelligence tests tend to enjoy better health and live longer as a result, too.

Neuroscientists have also identified some of the anatomical differences that might account for greater general intelligence.[23] The bark-like cerebral cortex is thicker and more wrinkled in more intelligent people, for example, and these also tend to have bigger brains overall.[24] And the long-distance neural connections linking different brain regions (called "white matter," since they are coated in a fatty sheath) appear to be wired differently too, forging more efficient networks for the transmission of signals.[25] Together, these differences may contribute to faster processing and greater short-term and long-term memory capacity that should make it easier to see patterns and process complex information.

It would be foolish to deny the value of these results and the undoubtedly important role that intelligence plays in our lives. The problems come when we place too much faith in those measures' capacity to represent someone's total intellectual potential[26] without recognizing the variation in behavior and performance that cannot be accounted for by these scores.[27]

If you consider surveys of lawyers, accountants, or engineers, for instance, the average IQ may lie around 125—showing that intelligence does give you an advantage. But the scores cover a considerable range, between around 95 (below average) and 157 (Termite territory).[28] And when you compare the individuals' success in those professions, those different scores can, at the very most, account for around 29 percent of the variance in performance, as measured by managers' ratings.[29] That is certainly a very significant chunk, but even if you take into account factors such as motivation, it still leaves a vast range in performance that cannot be accounted for by their intelligence.[30]

For any career, there are plenty of people of lower IQ who outperform those with much higher scores, and people with greater intelligence who don't make the most of their brainpower, confirming that qualities such as creativity or wise professional judgment just can't be accounted for by that one number alone. "It's a bit like being tall and playing basketball," David Perkins of the Harvard Graduate School of Education told me. If you don't meet a very basic threshold, you won't get far, but beyond that point other factors take over, he says.

Binet had warned us of this fact, and if you look closely at the data, this was apparent in the lives of the Termites. As a group, they were quite a bit more successful than the average American, but a vast number did not manage to fulfill their ambitions. The psychologist David Henry Feldman has examined the careers of the twenty-six brightest Termites, each of whom had a stratospheric IQ score of more than 180. Feldman was expecting to find each of these geniuses to have surpassed their peers, yet just four had reached a high level of professional distinction (becoming, for example, a judge or a highly honored architect); as a group, they were only slightly more successful than those scoring 30–40 points fewer.[31]

Consider Beatrice and Sara Ann—the two precocious young girls with IQs of 192 whom we met at the start of this chapter. Beatrice dreamed of being a sculptor and writer, but ended up dabbling in real estate with her husband's money—a stark contrast to the career of Oppenheimer, who had scored at the lower end of the group.[32] Sara Ann, meanwhile, earned a PhD, but apparently found it hard to concentrate on her career; by her fifties she was living a semi-nomadic life, moving from friend's house to friend's house, and briefly, in a commune. "I think I was made, as a child, to be far too self-conscious of my status as a 'Termite' . . . and given far too little to actually do with this mental endowment," she later wrote.[33]

We can't neglect the possibility that a few of the Termites may have made a conscious decision not to pursue a high-flying (and potentially stressful) career, but if general intelligence really were as important as Terman initially believed, you might have hoped for more of them to have reached great scientific, artistic, or political success.[34] "When we recall Terman's early optimism about his subjects' potential . . . there is the disappointing sense that they might have done more with their lives," Feldman concluded.

~

The interpretation of general intelligence as an all-powerful problem-solving-and-learning ability also has to contend with the Flynn Effect—a mysterious rise in IQ over the last few decades.

To find out more, I met Flynn at his son's house in Oxford, during a flying visit from his home in New Zealand.[35] Flynn is now a towering figure in intelligence research, but it was only meant to be a short distraction, he says: "I'm a moral philosopher who dabbles in psychology. And by dabbling I mean it's taken over half my time for the past thirty years."

Flynn's interest in IQ began when he came across troubling claims that certain racial groups are inherently less intelligent. He suspected that environmental effects would explain the differences in IQ scores: richer and more educated families will have a bigger vocabulary, for instance, meaning that their children perform better in the verbal parts of the test.

As he analyzed the various studies, however, he came across something even more puzzling: intelligence—for all races—appeared to have been rising over the decades. Psychologists had been slowly accounting for this by raising the bar of the exam—you had to answer more questions correctly to be given the same IQ score. But if you compare the raw data, the jump is remarkable, the equivalent of around thirty points over the last eighty years. "I thought, 'Why aren't psychologists dancing in the street over this? What the hell is going on?'" he told me.

Psychologists who believed that intelligence was largely inherited were dumbfounded. By comparing the IQ scores of siblings and strangers, they had estimated that genetics could explain around 70 percent of the variation between different people. But genetic evolution is slow: our genes could not possibly have changed quickly enough to produce the great gains in IQ score that Flynn was observing.

Flynn instead argues that we need to consider the large changes in society. Even though we are not schooled in IQ tests explicitly, we have been taught to see patterns and think in symbols and categories from a young age. Just think of the elementary school lessons that lead us to consider the different branches of the tree of life, the different elements and the forces of nature. The more children are exposed to these "scientific spectacles," the easier they find it to think in abstract terms more generally, Flynn suggests, leading to a steady rise in IQ over time. Our minds have been forged in Terman's image.[36]

Other psychologists were skeptical at first. But the Flynn Effect has been documented across Europe, Asia, the Middle East and South America (see below)—anywhere undergoing industrialization and Western-style educational reforms. The results suggest that general intelligence depends on the way our genes interact with the culture around us. Crucially—and in line with Flynn's theory of "scientific spectacles"—the scores in the different strands of the IQ test had not all risen equally. Non-verbal reasoning has improved much more than vocabulary or numerical reasoning, for instance—and other abilities that are not measured by IQ, like navigation, have actually deteriorated. We have simply refined a few specific skills that help us

The Flynn Effect by world region

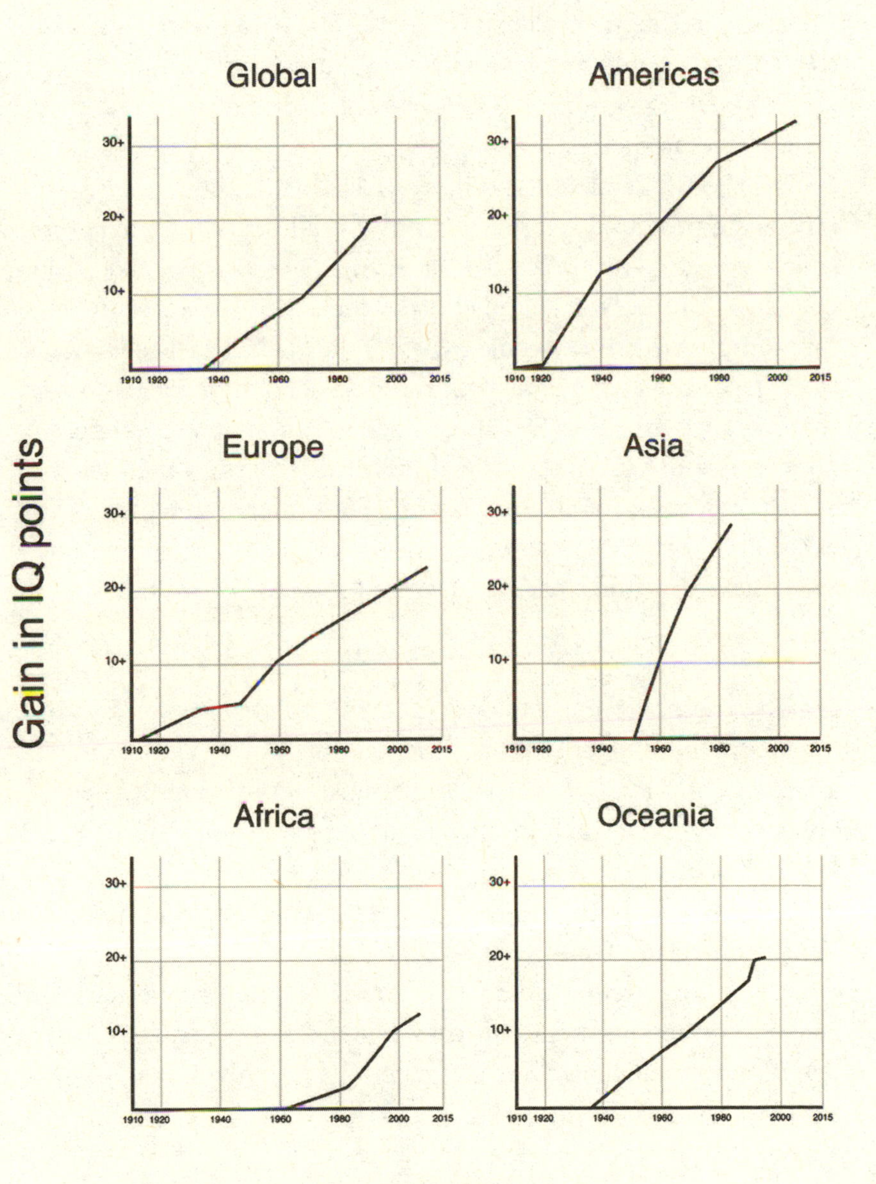

Source: OurWorldInData.org/Pietschnig, J., & Voracek, M. (2015).
One Century of Global IQ gains: A formal meta-analysis of the Flynn effect (1909–2013).
Perspectives on Psychological Science, 10(3), 282–306.

to think more abstractly. "Society makes highly different demands on us over time, and people have to respond." In this way, the Flynn Effect shows that we can't just train one type of reasoning and assume that *all* the useful problem-solving abilities that we have come to associate with greater intelligence will follow suit, as some theories would have predicted.[37]

This should be obvious from everyday life. If the rise of IQ really reflected a profound improvement in overall thinking, then even the smartest eighty-year-old (such as Flynn) would seem like a dunce compared to the average millennial. Nor do we see a rise in patents, for example, which you would expect if the skills measured by general intelligence tests were critical for the kind of technological innovation that Jess Oppenheimer had specialized in;[38] nor do we witness a preponderance of wise and rational political leaders, which you might expect if general intelligence alone were critical for truly insightful decision making. We do not live in the utopian future that Terman might have imagined, had he survived to see the Flynn Effect.[39]

~

Clearly, the skills measured by general intelligence tests are one important component of our mental machinery, governing how quickly we process and learn complex abstract information. But if we are to understand the full range of abilities in human decision making and problem solving, we need to expand our view to include many other elements—skills and styles of thinking that do not necessarily correlate strongly with IQ.

Attempts to define alternative forms of intelligence have often ended in disappointment, however. One popular buzzword has been "emotional intelligence," for instance.* It certainly makes sense that social skills determine many of our life outcomes, though critics have argued that some of the popular tests of "EQ" are flawed and fail

---

* Despite these criticisms, updated theories of emotional intelligence do prove to be critical for our understanding of intuitive reasoning and collective intelligence, as we will find out in Chapters 5 and 9.

to predict success better than IQ or measures of standard personality traits such as conscientiousness.[40]

In the 1980s, meanwhile, the psychologist Howard Gardner formulated a theory of "multiple intelligences" that featured eight traits, including interpersonal and intrapersonal intelligence, bodily-kinesthetic intelligence that makes you good at sports, and even "naturalistic intelligence"—whether you are good at discerning different plants in the garden or even whether you can tell the brand of car from the sound of its engine. But many researchers consider that Gardner's theory is too broad, without offering precise definitions and tests or any reliable evidence to support his conjectures, beyond the commonsense notion that some people do gravitate to some skills more than others.[41] After all, we've always known that some people are better at sports and others excel at music, but does that make them separate intelligences? "Why not also talk about stuffing-beans-up-your-nose intelligence?" Flynn said.

Robert Sternberg at Cornell University offers a middle ground with his Triarchic Theory of Successful Intelligence, which examines three particular types of intelligence—practical, analytical, and creative—that can together influence decision making in a diverse range of cultures and situations.[42]

When I called him one afternoon, he apologized for the sound of his young children playing in the garden outside. But he soon forgot the noise as he described his frustration with education today and the outdated tools we use to calculate mental worth.

He compares the lack of progress in intelligence testing to the enormous leaps made in other fields, like medicine: it is as if doctors were still using outdated nineteenth-century drugs to treat life-threatening disease. "We're at the level of using mercury to treat syphilis," he told me. "The SAT determines who gets into a good university, and then who gets into a good job—but all you get are good technicians with no common sense."

Like Terman before him, Sternberg's interest took root in childhood. Today, there is no questioning his brainpower: the American Psychological Association considered Sternberg the sixtieth most eminent psychologist in the twentieth century (twelve places above

Terman).[43] But as a second-grade child facing his first IQ test, his mind froze. When the results came in, it seemed clear to everyone—his teachers, his parents, and Sternberg himself—that he was a dunce. That low score soon became a self-fulfilling prophecy, and Sternberg is certain he would have continued on this downward spiral, had it not been for his teacher in the fourth grade.[44] "She thought there was more to a kid than an IQ score," he said. "My academic performance shot up just because she believed in me." It was only under her encouragement that his young mind began to flourish and blossom. Slippery concepts that had once slid from his grasp began to stick; he eventually became a first-class student.

As a freshman at Yale, he decided to take an introductory class in psychology to understand why he had been considered "so stupid" as a child—an interest that carried him to post-graduate research at Stanford, where he began to study developmental psychology. If IQ tests were so uninformative, he wondered, how could we better measure the skills that help people to succeed?

As luck would have it, observations of his own students started to provide the inspiration he needed. He remembers one girl, Alice, who had come to work in his lab. "Her test scores were terrific, she was a model student, but when she came in, she just didn't have any creative ideas," he said. She was the complete opposite of another girl, Barbara, whose scores had been good but not "spectacular," but who had been bursting with ideas to test in his lab.[45] Another, Celia, had neither the amazing grades of Alice, nor the brilliant ideas of Barbara, but she was incredibly pragmatic—she thought of exceptional ways to plan and execute experiments, to build an efficient team and to get her papers published.*

Inspired by Alice, Barbara, and Celia, Sternberg began to formulate a theory of human intelligence, which he defined as "the ability to achieve success in life, according to one's personal standards, within one's sociocultural context." Avoiding the (perhaps overly) broad definitions of Gardner's multiple intelligences, he confined his

---

* These case studies, often presented in Robert Sternberg's own writing, refer to real individuals. The names have been changed.

theory to those three abilities—analytical, creative, and practical—and considered how they might be defined, tested, and nurtured.

Analytical intelligence is essentially the kind of thinking that Terman was studying; it includes the abilities that allowed Alice to perform so well on her SATs. Creative intelligence, in contrast, examines our abilities "to invent, imagine, and suppose," as Sternberg puts it. While schools and universities already encourage this kind of thinking in creative writing classes, Sternberg points out that subjects such as history, science, and foreign languages can also incorporate exercises designed to measure and train creativity. A student looking at European history, for instance, might be asked, "Would the First World War have occurred, had Franz Ferdinand never been shot?" or, "What would the world look like today, if Germany had won the Second World War?" In a science lesson on animal vision, it might involve imagining a scene from the eyes of a bee. "Describe what a bee can see, that you cannot."[46]

Responding to these questions, students would still have a chance to show off their factual knowledge, but they are also being forced to exercise counterfactual thinking, to imagine events that have never happened—skills that are clearly useful in many creative professions. Jess Oppenheimer exercised this kind of thinking in his script writing and also his technical direction.

Practical intelligence, meanwhile, concerns a different kind of innovation: the ability to plan and execute an idea, and to overcome life's messy, ill-defined problems in the most pragmatic way possible. It includes traits like "metacognition"—whether you can judge your strengths and your weaknesses and work out the best ways to overcome them, and the unspoken, tacit knowledge that comes from experience and allows you to solve problems on the fly. It also includes some of the skills that others have called emotional or social intelligence—the ability to read motives and to persuade others to do what you want. Among the Termites, Shelley Smith Mydans's quick thinking as a war reporter, and her ability to navigate her escape from a Japanese prison camp, may best personify this kind of intelligence.

Of the three styles of thinking, practical intelligence may be the hardest to test or teach explicitly, but Sternberg suggests there are ways to cultivate it at school and university. In a business studies course, this may involve rating different strategies to deal with a personnel shortage;[47] in a history lesson on slavery, you might ask a student to consider the challenges of implementing the underground railroad for escaped slaves.[48] Whatever the subject, the core idea is to demand that students think of pragmatic solutions to an issue they may not have encountered before.

Crucially, Sternberg has since managed to test his theories in many diverse situations. At Yale University, for example, he helped set up a psychology summer program aimed at gifted high-school students. The children were tested according to his different measures of intelligence, and then divided randomly into groups and taught according to the principles of a particular kind of intelligence. After a morning studying the psychology of depression, for instance, some were asked to formulate their own theories based on what they had learned—a task to train creative intelligence; others were asked how they might apply that knowledge to help a friend who was suffering from mental illness—a task to encourage practical thinking. "The idea was that some kids will be capitalizing on their strengths, and others will be correcting their weaknesses," Sternberg told me.

The results were encouraging. They showed that teaching the children according to their particular type of intelligence improved their overall scores in a final exam—suggesting that education in general should help cater for people with a more creative or practical style of thinking. Moreover, Sternberg found that the practical and creative intelligence tests had managed to identify a far greater range of students from different ethnic and economic backgrounds—a refreshing diversity that was apparent as soon as they arrived for the course, Sternberg said.

In a later study, Sternberg recruited 110 schools (with more than 7,700 students in total) to apply the same principles to the teaching of mathematics, science, and English language. Again, the results were unequivocal—the children taught to develop their practical and creative intelligence showed greater gains overall, and even performed better on analytical, memory-based questions—suggesting that the more rounded approach had generally helped them to absorb and engage with the material.

Perhaps most convincingly, Sternberg's Rainbow Project collaborated with the admissions departments of various universities—including Yale, Brigham Young, and the University of California, Irvine—to build an alternative entrance exam that combines traditional SAT scores with measures of practical and creative intelligence. He found that the new test was roughly twice as accurate at predicting the students' GPA (grade point average) scores in their first year at university, compared to their SAT scores alone, which suggests that it does indeed capture different ways of thinking and reasoning that are valuable for success in advanced education.[49]

Away from academia, Sternberg has also developed tests of practical intelligence for business, and trialled them in executives and salespeople across industries, from local estate agents to Fortune 500 companies. One question asked the participants to rank potential approaches to different situations, such as how to deal with a perfectionist colleague whose slow progress may prevent your group from meeting its target, using various nudge techniques. Another scenario got them to explain how they would change their sales strategy when stocks are running low.

In each case, the questions test people's ability to prioritize tasks and weigh up the value of different options, to recognize the consequences of their actions and preempt potential challenges, and to persuade colleagues of pragmatic compromises that are necessary to keep a project moving without a stalemate. Crucially, Sternberg has found that these tests predicted measures of success such as yearly profits, the chances of winning a professional award, and overall job satisfaction.

In the military, meanwhile, Sternberg examined various measures of leadership performance among platoon commanders, company commanders, and battalion commanders. They were asked how to deal with soldier insubordination, for instance—or the best way to communicate the goals of a mission. Again, practical intelligence—and tacit knowledge, in particular—predicted their leadership ability better than traditional measures of general intelligence.[50]

Sternberg's measures may lack the elegance of a one-size-fits-all

IQ score, but they are a step closer to measuring the kind of thinking that allowed Jess Oppenheimer and Shelley Smith Mydans to succeed where other Termites failed.[51] "Sternberg's on the right track," Flynn told me. "He was excellent in terms of showing that it was possible to measure more than analytic skills."

Disappointingly, acceptance has been slow. Although his measures have been adopted at Tufts University and Oklahoma State University, they are still not widespread. "People may say things will change, but then things go back to the way they were before," Sternberg said. Just like when he was a boy, teachers are still too quick to judge a child's potential based on narrow, abstract tests—a fact he has witnessed in the education of his own children, one of whom is now a successful Silicon Valley entrepreneur. "I have five kids and all of them at one time or another have been diagnosed as potential losers," he said, "and they've done fine."

~

While Sternberg's research may not have revolutionized education in the way he had hoped, it has inspired other researchers to build on his concept of tacit knowledge—including some intriguing new research on the concept of "cultural intelligence."

Soon Ang, a professor of management at the Nanyang Technological University in Singapore, has pioneered much of this work. In the late 1990s, she was acting as a consultant to several multinational companies who asked her to pull together a team of programmers, from many different countries, to help them cope with the "Y2K bug."

The programmers were undeniably intelligent and experienced, but Ang observed that they were disappointingly ineffective at working together: she found that Indian and Filipino programmers would appear to agree on a solution to a problem, for instance, only for the members to then implement it in different and incompatible ways. Although the team members were all speaking the same language, Ang realized that they were struggling to bridge the cultural divide and comprehend the different ways of working.

Inspired, in part, by Robert Sternberg's work, she developed a measure of "cultural intelligence" (CQ) that examines your general sensitivity to different cultural norms. As one simple example: a Brit or American may be surprised to present an idea to Japanese colleagues, only to be met with silence. Someone with low cultural intelligence may interpret the reaction as a sign of disinterest; someone with high cultural intelligence would realize that, in Japan, you may need to explicitly ask for feedback before getting a response—even if the reaction is positive. Or consider the role of small talk in building a relationship. In some European countries, it's much better to move directly to the matter at hand, but in India it is important to take the time to build relationships—and someone with high cultural intelligence would recognize that fact.

Ang found that some people are consistently better at interpreting those signs than others. Importantly, the measures of cultural intelligence test not only your knowledge of a specific culture, but also your general sensitivity to the potential areas of misunderstanding in unfamiliar countries, and how well you would adapt to them. And like Sternberg's measures of practical intelligence, these tacit skills don't correlate very strongly with IQ or other tests of academic potential—reaffirming the idea that they are measuring different things. As Ang's programmers had shown, you could have high general intelligence but low cultural intelligence.

"CQ" has now been linked to many measures of success. It can predict how quickly expats will adapt to their new life, the performance of international salespeople, and participants' abilities to negotiate.[52] Beyond business, cultural intelligence may also determine the experiences of students studying abroad, charity workers in disaster zones, and teachers at international schools—or even your simple enjoyment of a holiday abroad.

~

My conversations with Flynn and Sternberg were humbling. Despite having performed well academically, I have to admit that I lack many of the other skills that Sternberg's tests have been measuring, including many forms of tacit knowledge that may be obvious to some people.

Imagine, for instance, that your boss is a micromanager and wants to have the last say on every project—a problem many of us will have encountered. Having spoken to Sternberg, I realized that someone with practical intelligence might skilfully massage the micromanager's sense of self-importance by suggesting two solutions to a problem: the preferred answer, and a decoy they could reject while feeling they have still left their mark on the project. It's a strategy that had never once occurred to me.

Or consider you are a teacher, and you find a group of children squabbling in the playground. Do you scold them, or do you come up with a simple distraction that will cause them to forget their quarrel? To my friend Emma, who teaches in a primary school in Oxford, the latter is second nature; her mind is full of games and subtle hints to nudge their behavior. But when I tried to help her out in the classroom one day, I was clueless, and the children were soon running rings around me.

I'm not unusual in this. In Sternberg's tests of practical intelligence, a surprising number of people lacked this pragmatic judgment, even if, like me, they score higher than average on other measures of intelligence, and even if they had years of experience in the job at hand. The studies do not agree on the exact relation, though. At best, the measures of tacit knowledge are very modestly linked to IQ scores; at worst, they are negatively correlated. Some people just seem to find it easier to implicitly learn the rules of pragmatic problem solving—and that ability is not very closely related to general intelligence.

For our purposes, it's also worth paying special attention to counterfactual thinking—an element of creative intelligence that allows us to think of the alternative outcomes of an event or to momentarily imagine ourselves in a different situation. It's the capacity to ask "what if . . . ?" and without it, you may find yourself helpless when faced with an unexpected challenge. Without being able to reappraise your past, you'll also struggle to learn from your mistakes to find better solutions in the future. Again, that's neglected on most academic tests.

In this way, Sternberg's theories help us to understand the frustrations of intelligent people who somehow struggle with some of the basic tasks of working life—such as planning projects, imagining the consequences of their actions, and preempting problems before they emerge. Failed entrepreneurs may be one example: around nine out of ten new business ventures fail, often because the innovator has found a good idea but lacks the capacity to deal with the challenges of implementing it.

If we consider that SATs or IQ tests reflect a unitary, underlying mental energy—a "raw brainpower"—that governs all kinds of problem solving, this behavior doesn't make much sense; people of high general intelligence should have picked up those skills. Sternberg's theory allows us to disentangle those other components and then define and measure them with scientific rigor, showing that they are largely independent abilities.

These are important first steps in helping us to understand why apparently clever people may lack the good judgment that we might have expected given their academic credentials. This is just the start, however. In the next chapters we will discover many other essential thinking styles and cognitive skills that had been neglected by psychologists—and the reasons that greater intelligence, rather than protecting us from error, can sometimes drive us to make even bigger mistakes. Sternberg's theories only begin to scratch the surface.

~

In hindsight, Lewis Terman's own life exemplifies many of these findings. From early childhood he had always excelled academically, rising from his humble background to become president of the American Psychological Association. Nor should we forget the fact that he masterminded one of the first and most ambitious cohort studies ever conducted, collecting reams of data that scientists continued to study four decades after his death. He was clearly a highly innovative man.

And yet it is now so easy to find glaring flaws in his thinking. A good scientist should leave no stone uncovered before reaching a conclusion—but Terman turned a blind eye to data that might have contradicted his own preconceptions. He was so sure of the genetic nature of intelligence that he neglected to hunt for talented children in poorer neighborhoods. And he must have known that meddling in his subjects' lives would skew the results, but he often offered financial support and professional recommendations to his Termites, boosting their chances of success. He was neglecting the most basic (tacit) knowledge of the scientific method, which even the most inexperienced undergraduate should take for granted.

This is not to mention his troubling political leanings. Terman's interest in social engineering led him to join the Human Betterment Foundation—a group that called for the compulsory sterilization of those showing undesirable qualities.[53] Moreover, when reading Terman's early papers, it is shocking how easily he dismissed the intellectual potential of African Americans and Hispanics, based on a mere handful of case studies. Describing the poor scores of just two Portuguese boys, he wrote: "Their dullness seems to be racial, or at least inherent in the family stocks from which they came."[54] Further research, he was sure, would reveal "enormously significant racial differences in general intelligence."

Perhaps it is unfair to judge the man by today's standards; certainly, some psychologists believe that we should be kind to Terman's faults, product as he was of a different time. Except that we know Terman had been exposed to other points of view; he must have read Binet's concerns about the misuse of his intelligence test.

A wiser man might have explored these criticisms, but when Terman was challenged on these points, he responded with knee-jerk vitriol rather than reasoned argument. In 1922, the journalist and political commentator Walter Lippmann wrote an article in the *New Republic*, questioning the IQ test's reliability. "It is not possible," Lippmann wrote, "to imagine a more contemptible proceeding than to confront a child with a set of puzzles, and after an hour's monkeying with them, proclaim to the child, or to his parents, that here is a C– individual."[55]

Lippmann's skepticism was entirely understandable, yet Terman's response was an *ad hominem* attack: "Now it is evident that Mr Lippmann has been seeing red; also, that seeing red is not very conducive to seeing clearly," he wrote in response. "Clearly, something has hit the bulls-eye of one of Mr Lippmann's emotional complexes."[56]

Even the Termites had started to question the values of their test results by the ends of their lives. Sara Ann—the charming little girl with an IQ of 192, who had "bribed" her experimenters with a gumdrop—certainly resented the fact that she had not cultivated other cognitive skills that had not been measured in her test. "My great regret is that my left-brain parents, spurred on by my Terman group experience, pretty completely bypassed any encouragement of whatever creative talent I may have had," she wrote. "I now see the latter area as of greater significance, and intelligence as its hand-maiden. [I'm] sorry I didn't become aware of this fifty years ago."[57]

Terman's views softened slightly over the years, and he would later admit that "intellect and achievement are far from perfectly correlated," yet his test scores continued to dominate his opinions of the people around him; they even cast a shadow over his relationships with his family. According to Terman's biographer, Henry Minton, each of his children and grandchildren had taken the IQ test, and his love for them appeared to vary according to the results. His letters were full of pride for his son, Fred, a talented engineer and an early pioneer in Silicon Valley; his daughter, Helen, barely merited a mention.

Perhaps most telling are his granddaughter Doris's recollections of family dinners, during which the place settings were arranged in order of intelligence: Fred sat at the head of the table next to Lewis; Helen and her daughter Doris sat at the other end, where they could help the maid.[58] Each family member placed according to a test they had taken years before—a tiny glimpse, perhaps, of the way Terman would have liked to arrange us all.

2

Entangled arguments: *The dangers of "dysrationalia"*

It is June 17, 1922, and two middle-aged men—one short and squat, the other tall and lumbering with a walrus moustache—are sitting on the beach in Atlantic City, New Jersey. They are Harry Houdini and Arthur Conan Doyle[1]—and by the end of the evening, their friendship will never be the same again.

It ended as it began—with a séance. Spiritualism was all the rage among London's wealthy elite, and Conan Doyle was a firm believer, attending five or six gatherings a week. He even claimed that his wife Jean had some psychic talent, and that she had started to channel a spirit guide, Phineas, who dictated where they should live and when they should travel.

Houdini, in contrast, was a skeptic, but he still claimed to have an open mind, and on a visit to England two years previously, he had contacted Conan Doyle to discuss his recent book on the subject. Despite their differences, the two men had quickly struck up a fragile friendship and Houdini had even agreed to visit Conan Doyle's favorite medium, who claimed to channel ectoplasm through her mouth and vagina; he quickly dismissed her powers as simple stage magic. (I'll spare you the details.)

Now Conan Doyle was in the middle of an American book tour, and he invited Houdini to join him in Atlantic City.

The visit had begun amicably enough. Houdini had helped to teach Conan Doyle's boys to dive, and the group were resting at the seafront when Conan Doyle decided to invite Houdini up to his hotel room for an impromptu séance, with Jean as the medium. He knew that Houdini had been mourning the loss of his mother, and he hoped that his wife might be able to make contact with the other side.

And so they returned to the Ambassador Hotel, closed the curtains, and waited for inspiration to strike. Jean sat in a kind of trance with a pencil in one hand as the men sat by and watched. She then began to strike the table violently with her hands—a sign that the spirit had descended.

"Do you believe in God?" she asked the spirit, who responded by moving her hand to knock again on the table. "Then I shall make the sign of the cross."

She sat with her pen poised over the writing pad, before her hand began to fly wildly across the page.

"Oh, my darling, thank God, at last I'm through," the spirit wrote. "I've tried oh so often—now I am happy. Why, of course, I want to talk to my boy—my own beloved boy. Friends, thank you, with all my heart for this—you have answered the cry of my heart—and of his—God bless him."

By the end of the séance, Jean had written around twenty pages in "angular, erratic script." Her husband was utterly bewitched. "It was a singular scene—my wife with her hand flying wildly, beating the table while she scribbled at a furious rate, I sitting opposite and tearing sheet after sheet from the block as it was filled up."

Houdini, in contrast, cut through the charade with a number of questions. Why had his mother, a Jew, professed herself to be a Christian? How had this Hungarian immigrant written her messages in perfect English—"a language which she had never learned!"? And why did she not bother to mention that it was her birthday?

Houdini later wrote about his skepticism in an article for the *New York Sun*. It was the start of an increasingly public dispute between the two men, and their friendship never recovered before the escapologist's death four years later.[2]

Even then, Conan Doyle could not let the matter rest. Egged on, perhaps, by his "spirit guide" Phineas, he attempted to address and dismiss all of Houdini's doubts in an article for *The Strand* magazine. His reasoning was more fanciful than any of his fictional works, not least in claiming that Houdini himself was in command of a "dematerializing and reconstructing force" that allowed him to slip in and out of chains.

"Is it possible for a man to be a very powerful medium all his life, to use that power continually, and yet never to realize that the gifts he is using are those which the world calls mediumship?" he wrote. "If that be indeed possible, then we have a solution of the Houdini enigma."

~

Meeting these two men for the first time, you would have been forgiven for expecting Conan Doyle to be the more critical thinker. A doctor of medicine and a best-selling writer, he exemplified the abstract reasoning that Terman was just beginning to measure with his intelligence tests. Yet it was the professional illusionist, a Hungarian immigrant whose education had ended at the age of twelve, who could see through the fraud.

Some commentators have wondered whether Conan Doyle was suffering from a form of madness. But let's not forget that many of his contemporaries believed in spiritualism—including scientists such as the physicist Oliver Lodge, whose work on electromagnetism brought us the radio, and the naturalist Alfred Russel Wallace, a contemporary of Charles Darwin who had independently conceived the theory of natural selection. Both were formidable intellectual figures, but they remained blind to any evidence debunking the paranormal.

We've already seen how our definition of intelligence could be expanded to include practical and creative reasoning. But those theories do not explicitly examine our *rationality*, defined as our capacity to make the *optimal* decisions needed to meet our goals, given the resources we have to hand, and to form beliefs based on evidence, logic, and sound reasoning.*

---

* Cognitive scientists such as Keith Stanovich describe two classes of rationality. Instrumental rationality is defined as "the optimization of someone's goal fulfillment," or, less technically, as "behaving so that you get exactly what you want, given the resources available to you." Epistemic rationality, meanwhile, concerns "how well your beliefs map onto the actual structure of the world." By falling for fraudulent mediums, Conan Doyle was clearly lacking in the latter.

While decades of psychological research have documented humanity's more irrational tendencies, it is only relatively recently that scientists have started to measure how that irrationality varies between individuals, and whether that variance is related to measures of intelligence. They are finding that the two are far from perfectly correlated: it is possible to have a very high SAT score that demonstrates good abstract thinking, for instance, while still performing badly on these new tests of rationality—a mismatch known as "dysrationalia."

Conan Doyle's life story—and his friendship with Houdini, in particular—offers the perfect lens through which to view this cutting-edge research.[3] I certainly wouldn't claim that any kind of faith is inherently irrational, but I am interested in the fact that fraudsters were able to exploit Conan Doyle's beliefs to fool him time after time. He was simply blind to the evidence, including Houdini's testimonies. Whatever your views on paranormal belief in general, he did not need to be quite so gullible at such great personal cost.

Conan Doyle is particularly fascinating because we know, through his writing, that he was perfectly aware of the laws of logical deduction. Indeed, he started to dabble in spiritualism *at the same time* that he first created Sherlock Holmes:[4] he was dreaming up literature's greatest scientific mind during the day, but failed to apply those skills of deduction at night. If anything, his intelligence seems to have only allowed him to come up with increasingly creative arguments to dismiss the skeptics and justify his beliefs; he was bound more tightly than Houdini in his chains.

Besides Doyle, many other influential thinkers of the last hundred years may have also been afflicted by this form of the intelligence trap. Even Einstein—whose theories are often taken to be the pinnacle of human intelligence—may have suffered from this blinkered reasoning, leading him to waste the last twenty-five years of his career with a string of embarrassing failures.

Whatever your specific situation and interests, this research will explain why so many of us make mistakes that are blindingly obvious to all those around us—and continue to make those errors long after the facts have become apparent.

Houdini himself seems to have intuitively understood the vulnerability of the intelligent mind. "As a rule, I have found that the greater brain a man has, and the better he is educated, the easier it has been to mystify him," he once told Conan Doyle.[5]

~

A true recognition of dysrationalia—and its potential for harm—has taken decades to blossom, but the roots of the idea can be found in the now legendary work of two Israeli researchers, Daniel Kahneman and Amos Tversky, who identified many cognitive biases and heuristics (quick-and-easy rules of thumb) that can skew our reasoning.

One of their most striking experiments asked participants to spin a "wheel of fortune," which landed on a number between 1 and 100, before considering general knowledge questions—such as estimating the number of African countries that are represented in the UN. The wheel of fortune should, of course, have had no influence on their answers—but the effect was quite profound. The lower the quantity on the wheel, the smaller their estimate—the arbitrary value had planted a figure in their mind, "anchoring" their judgment.[6]

You have probably fallen for anchoring yourself many times while shopping during sales. Suppose you are looking for a new TV. You had expected to pay around $150, but then you find a real bargain: a $300 item reduced to $200. Seeing the original price anchors your perception of what is an acceptable price to pay, meaning that you will go above your initial budget. If, on the other hand, you had not seen the original price, you would have probably considered it too expensive, and moved on.

You may also have been prey to the availability heuristic, which causes us to overestimate certain risks based on how easily the dangers come to mind, thanks to their vividness. It's the reason that many people are more worried about flying than driving—because reports of plane crashes are often so much more emotive, despite the fact that it is actually far more dangerous to step into a car.

There is also framing: the fact that you may change your opinion based on the way information is phrased. Suppose you are considering a medical treatment for 600 people with a deadly illness and it has a 1 in 3 success rate. You can be told either that "200 people will be saved using this treatment" (the gain framing) or that "400 people will die using this treatment" (the loss framing). The statements mean exactly the same thing, but people are more likely to endorse the statement when it is presented in the gain framing; they passively accept the facts as they are given to them without thinking what they really mean. Advertisers have long known this: it's the reason that we are told that foods are 95 percent fat free (rather than being told they are "5 percent fat").

Other notable biases include the sunk cost fallacy (our reluctance to give up on a failing investment even if we will lose more trying to sustain it), and the gambler's fallacy—the belief that if the roulette wheel has landed on black, it's more likely the next time to land on red. The probability, of course, stays exactly the same. An extreme case of the gambler's fallacy is said to have been observed in Monte Carlo in 1913, when the roulette wheel fell twenty-six times on black—and the visitors lost millions as the bets on red escalated. But it is not just witnessed in casinos; it may also influence family planning. Many parents falsely believe that if they have already produced a line of sons, then a daughter is more likely to come next. With this logic, they may end up with a whole football team of boys.

Given these findings, many cognitive scientists divide our thinking into two categories: "system 1," intuitive, automatic, "fast thinking" that may be prey to unconscious biases; and "system 2," "slow," more analytical, deliberative thinking. According to this view—called dual-process theory—many of our irrational decisions come when we rely too heavily on system 1, allowing those biases to muddy our judgment.

Yet none of the early studies by Kahneman and Tversky had tested whether our irrationality varies from person to person. Are some people more susceptible to these biases, while others are immune, for instance? And how do those tendencies relate to our general intelligence? Conan Doyle's story is surprising because we intuitively expect more intelligent people, with their greater analytical minds, to act more rationally—but as Tversky and Kahneman had shown, our intuitions can be deceptive.

If we want to understand why smart people do stupid things, these are vital questions.

During a sabbatical at the University of Cambridge in 1991, a Canadian psychologist called Keith Stanovich decided to address these issues head on. With a wife specializing in learning difficulties, he had long been interested in the ways that some mental abilities may lag behind others, and he suspected that rationality would be no different. The result was an influential paper introducing the idea of dysrationalia as a direct parallel to other disorders like dyslexia and dyscalculia.

It was a provocative concept—aimed as a nudge in the ribs to all the researchers examining bias. "I wanted to jolt the field into realizing that it had been ignoring individual differences," Stanovich told me.

Stanovich emphasises that dysrationalia is not just limited to system 1 thinking. Even if we are reflective enough to detect when our intuitions are wrong, and override them, we may fail to use the right "mindware"—the knowledge and attitudes that should allow us to reason correctly.[7] If you grow up among people who distrust scientists, for instance, you may develop a tendency to ignore empirical evidence, while putting your faith in unproven theories.[8] Greater intelligence wouldn't necessarily stop you forming those attitudes in the first place, and it is even possible that your greater capacity for learning might then cause you to accumulate more and more "facts" to support your views.[9]

Circumstantial evidence would suggest that dysrationalia is common. One study of the high-IQ society Mensa, for example, showed that 44 percent of its members believed in astrology, and 56 percent believed that the Earth had been visited by extra-terrestrials.[10] But rigorous experiments, specifically exploring the link between intelligence and rationality, were lacking.

Stanovich has now spent more than two decades building on those foundations with a series of carefully controlled experiments.

To understand his results, we need some basic statistical theory. In psychology and other sciences, the relationship between two variables is usually expressed as a correlation coefficient between 0 and 1. A perfect correlation would have a value of 1—the two parameters would essentially be measuring the same thing; this is unrealistic for most studies of human health and behavior (which are determined by so many variables), but many scientists would consider a "moderate" correlation to lie between 0.4 and 0.59.[11]

Using these measures, Stanovich found that the relationships between rationality and intelligence were generally very weak. SAT scores revealed a correlation of just 0.1 and 0.19 with measures of the framing bias and anchoring, for instance.[12] Intelligence also appeared to play only a tiny role in the question of whether we are willing to delay immediate gratification for a greater reward in the future—a tendency known as "temporal discounting." In one test, the correlation with SAT scores was as small as 0.02. That's an extraordinarily modest correlation for a trait that many might assume comes hand in hand with a greater analytical mind. The sunk cost bias also showed almost no relationship to SAT scores in another study.[13]

Gui Xue and colleagues at Beijing Normal University, meanwhile, have followed Stanovich's lead, finding that the gambler's fallacy is actually a little more common among the more academically successful participants in his sample.[14] That's worth remembering: when playing roulette, don't think you are smarter than the wheel.

Even trained philosophers are vulnerable. Participants with PhDs in philosophy are just as likely to suffer from framing effects, for example, as everyone else—despite the fact that they should have been schooled in logical reasoning.[15]

You might at least expect that more intelligent people could learn to recognize these flaws. In reality, most people assume that they are less vulnerable than other people, and this is equally true of the "smarter" participants. Indeed, in one set of experiments studying some of the classic cognitive biases, Stanovich found that people with higher SAT scores actually had a slightly larger "bias blind spot" than people who were less academically gifted.[16] "Adults with more cognitive ability are aware of their intellectual status and expect to outperform others on most cognitive tasks," Stanovich told me. "Because these cognitive biases are presented to them as essentially cognitive tasks, they expect to outperform on them as well."

From my interactions with Stanovich, I get the impression that he is extremely cautious about promoting his findings, meaning he has not achieved the same kind of fame as Daniel Kahneman, say—but colleagues within his field believe that these theories could be truly game-changing. "The work he has done is some of the most important research in cognitive psychology—but it's sometimes underappreciated," agreed Gordon Pennycook, a professor at the University of Regina, Canada, who has also specialized in exploring human rationality.

Stanovich has now refined and combined many of these measures into a single test, which is informally called the "rationality quotient." He emphasises that he does not wish to devalue intelligence tests—they "work quite well for what they do"—but to improve our understanding of these other cognitive skills that may also determine our decision making, and place them on an equal footing with the existing measures of cognitive ability.

"Our goal has always been to give the concept of rationality a fair hearing—almost as if it had been proposed prior to intelligence," he wrote in his scholarly book on the subject.[17] It is, he says, a "great irony" that the thinking skills explored in Kahneman's Nobel Prize-winning work are still neglected in our most well-known assessment of cognitive ability.[18]

After years of careful development and verification of the various sub-tests, the first iteration of the "Comprehensive Assessment of Rational Thinking" was published at the end of 2016. Besides measures of the common cognitive biases and heuristics, it also included probabilistic and statistical reasoning skills—such as the ability to assess risk—that could improve our rationality, and questionnaires concerning contaminated mindware such as anti-science attitudes.

For a taster, consider the following question, which aims to test the "belief bias." Your task is to consider whether the conclusion follows logically, based *only* on the opening two premises.

All living things need water.
Roses need water.
Therefore, roses are living things.

What did you answer? According to Stanovich's work, 70 percent of university students believe that this is a valid argument. But it isn't, since the first premise only says that "all living things need water"—not that "all things that need water are living."

If you still struggle to understand why that makes sense, compare it to the following statements:

All insects need oxygen.
Mice need oxygen.
Therefore mice are insects.

The logic of the two statements is exactly the same—but it is far easier to notice the flaw in the reasoning when the conclusion clashes with your existing knowledge. In the first example, however, you have to put aside your preconceptions and think, carefully and critically, about the specific statements at hand—to avoid thinking that the argument is right just because the conclusion makes sense with what you already know.[19] That's an important skill whenever you need to appraise a new claim.

When combining all these sub-tests, Stanovich found that the overall correlation with measures of general intelligence, such as SAT scores, was modest: around 0.47 on one test. Some overlap was to be expected, especially given the fact that several of these measures, such as probabilistic reasoning, would be aided by mathematical ability and other aspects of cognition measured by IQ tests and SATs. "But that still leaves enough room for the discrepancies between rationality and intelligence that lead to smart people acting foolishly," Stanovich said.

With further development, the rationality quotient could be used in recruitment to assess the quality of a potential employee's decision making; Stanovich told me that he has already had significant interest from law firms, financial institutions, and executive headhunters.

Stanovich hopes his test may also be a useful tool to assess how students' reasoning changes over a school or university course. "This, to me, would be one of the more exciting uses," Stanovich said. With that data, you could then investigate which interventions are most successful at cultivating more rational thinking styles.

~

While we wait to see that work in action, cynics may question whether RQ really does reflect our behavior in real life. After all, the IQ test is sometimes accused of being too abstract. Is RQ—based on artificial, imagined scenarios—any different?

Some initial answers come from the work of Wändi Bruine de Bruin at Leeds University. Inspired by Stanovich's research, her team first designed their own scale of "adult decision-making competence," consisting of seven tasks measuring biases like framing, measures of risk perception, and the tendency to fall for the sunk cost fallacy (whether you are likely to continue with a bad investment or not). The team also examined overconfidence by asking the subjects some general knowledge questions, and then asking them to gauge how sure they were that each answer was correct.

Unlike many psychological studies, which tend to use university students as guinea pigs, Bruine de Bruin's experiment examined a diverse sample of people, aged eighteen to eighty-eight, with a range of educational backgrounds—allowing her to be sure that any results reflected the population as a whole.

As Stanovich has found with his tests, the participants' decision-making skills were only moderately linked to their intelligence; academic success did not necessarily make them more rational decision makers.

But Bruine de Bruin then decided to see how both measures were related to their behaviors in the real world. To do so, she asked participants to declare how often they had experienced various stressful life events, from the relatively trivial (such as getting sunburnt or missing a flight), to the serious (catching an STD or cheating on your partner) and the downright awful (being put in jail).[20] Although the measures of general intelligence did seem to have a small effect on these outcomes, the participants' rationality scores were about three times more important in determining their behavior.

These tests clearly capture a more general tendency to be a careful, considered thinker that was not reflected in more standard measures of cognitive ability; you can be intelligent and irrational—as Stanovich had found—and this has serious consequences for your life.

Bruine de Bruin's findings can offer us some insights into other peculiar habits of intelligent people. One study from the London School of Economics, published in 2010, found that people with higher IQs tend to consume more alcohol and may be more likely to smoke or take illegal drugs, for instance—supporting the idea that intelligence does not necessarily help us to weigh up short-term benefits against the long-term consequences.[21]

People with high IQs are also just as likely to face financial distress, such as missing mortgage payments, bankruptcy, or credit card debt. Around 14 percent of people with an IQ of 140 had reached their credit limit, compared to 8.3 percent of people with an average IQ of 100. Nor were they any more likely to put money away in long-term investments or savings; their accumulated wealth each year was just a tiny fraction greater. These facts are particularly surprising, given that more intelligent (and better educated) people do tend to have more stable jobs with higher salaries, which suggests that their financial distress is a consequence of their decision making, rather than, say, a simple lack of earning power.[22]

The researchers suggested that more intelligent people veer close to the "financial precipice" in the belief that they will be better able to deal with the consequences afterward. Whatever the reason, the results suggest that smarter people are not investing their money in the more rational manner that economists might anticipate; it is another sign that intelligence does not necessarily lead to better decision making.

~

As one vivid example, consider the story of Paul Frampton. A brilliant physicist at the University of North Carolina, his work ranged from a new theory of dark matter (the mysterious, invisible mass holding our universe together) to the prediction of a subatomic particle called the "axigluon," a theory that is inspiring experiments at the Large Hadron Collider.

In 2011, however, he began online dating, and soon struck up a friendship with a former bikini model named Denise Milani. In January the next year, she invited him to visit her on a photoshoot in La Paz, Bolivia. When he arrived, however, he found a message—she'd had to leave for Argentina instead. But she'd left her bag. Could he pick it up and bring it to her?

Alas, he arrived in Argentina but there was still no sign of Milani. Losing patience, he decided to return to the US, where he checked in her suitcase with his own luggage. A few minutes later, an announcement called him to meet the airport staff at his gate. Unless you suffer from severe dysrationalia yourself, you can probably guess what happened next. He was subsequently charged with transporting two kilograms of cocaine.

Fraudsters, it turned out, had been posing as Milani—who really is a model, but knew nothing of the scheme and had never been in touch with Frampton. They would have presumably intercepted the bag once he had carried it over the border.

Frampton had been warned about the relationship. "I thought he was out of his mind, and I told him that," John Dixon, a fellow physicist and friend of Frampton's, said in the *New York Times*. "But he really believed that he had a pretty young woman who wanted to marry him."[23]

We can't really know what was going through Frampton's mind. Perhaps he suspected that "Milani" was involved in some kind of drug smuggling operation but thought that this was a way of proving himself to her. His love for her seems to have been real, though; he even tried to message her in prison, after the scam had been uncovered. For some reason, however, he just hadn't been able to weigh up the risks, and had allowed himself to be swayed by impulsive, wishful thinking.

~

If we return to that séance in Atlantic City, Arthur Conan Doyle's behavior would certainly seem to fit neatly with theories of dysrationalia, with compelling evidence that paranormal and superstitious beliefs are surprisingly common among the highly intelligent.

According to a survey of more than 1,200 participants, people with college degrees are just as likely to endorse the existence of UFOs, and they were even more credulous of extrasensory perception and "psychic healing" than people with a worse education.[24] (The education level here is an imperfect measure of intelligence, but it gives a general idea that the abstract thinking and knowledge required to enter university does not translate into more rational beliefs.)

Needless to say, all of the phenomena above have been repeatedly disproven by credible scientists—yet it seems that many smart people continue to hold on to them regardless. According to dual-process (fast/slow thinking) theories, this could just be down to cognitive miserliness. People who believe in the paranormal rely on their gut feelings and intuitions to think about the sources of their beliefs, rather than reasoning in an analytical, critical way.[25]

This may be true for many people with vaguer, less well-defined beliefs, but there are some particular elements of Conan Doyle's biography that suggest his behavior can't be explained quite so simply. Often, it seemed as if he was using analytical reasoning from system 2 to rationalize his opinions and dismiss the evidence. Rather than thinking *too little*, he was thinking *too much*.

Consider how Conan Doyle was once infamously fooled by two schoolgirls. In 1917—a few years before he met Houdini—sixteen-year-old Elsie Wright and nine-year-old Frances Griffith claimed to have photographed a population of fairies frolicking around a stream in Cottingley, West Yorkshire. Through a contact at the local Theosophical Society, the pictures eventually landed in Conan Doyle's hands.

Many of his acquaintances were highly skeptical, but he fell for the girls' story hook, line, and sinker.[26] "It is hard for the mind to grasp what the ultimate results may be if we have actually proved the existence upon the surface of this planet of a population which may be as numerous as the human race," he wrote in *The Coming of Fairies*.[27] In reality, they were cardboard cut-outs, taken from *Princess Mary's Giftbook*[28]—a volume that had also included some of Conan Doyle's own writing.[29]

What's fascinating is not so much the fact that he fell for the fairies in the first place, but the extraordinary lengths that he went to explain away any doubts. If you look at the photographs carefully, you can even see hatpins holding one of the cut-outs together. But where others saw pins, he saw the gnome's belly button—proof that fairies are linked to their mothers in the womb with an umbilical cord. Conan Doyle even tried to draw on modern scientific discoveries to explain the fairies' existence, turning to electromagnetic theory to claim that they were "constructed in material which threw out shorter or longer vibrations," rendering them invisible to humans.

As Ray Hyman, a professor of psychology at the University of Oregon, puts it: "Conan Doyle used his intelligence and cleverness to dismiss all counter-arguments. . . . [He] was able to use his smartness to outsmart himself."[30]

The use of system 2 "slow thinking" to rationalize our beliefs even when they are wrong leads us to uncover the most important and pervasive form of the intelligence trap, with many disastrous consequences; it can explain not only the foolish ideas of people such as Conan Doyle, but also the huge divides in political opinion about issues such as gun crime and climate change.

~

So what's the scientific evidence?

The first clues came from a series of classic studies from the 1970s and 1980s, when David Perkins of Harvard University asked students to consider a series of topical questions, such as: "Would a nuclear disarmament treaty reduce the likelihood of world war?" A truly rational thinker should consider both sides of the argument, but Perkins found that more intelligent students were no more likely to consider any alternative points of view. Someone in favor of nuclear disarmament, for instance, might not explore the issue of trust: whether we could be sure that all countries would honor the agreement. Instead, they had simply used their abstract reasoning skills and factual knowledge to offer more elaborate justifications of their own point of view.[31]

This tendency is sometimes called the confirmation bias, though several psychologists—including Perkins—prefer to use the more general term "myside bias" to describe the many different kinds of tactics we may use to support our viewpoint and diminish alternative opinions. Even student lawyers, who are explicitly trained to consider the other side of a legal dispute, performed very poorly.

Perkins later considered this to be one of his most important discoveries.[32] "Thinking about the other side of the case is a perfect example of a good reasoning practice," he said. "Why, then, do student lawyers with high IQs and training in reasoning that includes anticipating the arguments of the opposition prove to be as subject to confirmation bias or myside bias, as it has been called, as anyone else? To ask such a question is to raise fundamental issues about conceptions of intelligence."[33]

Later studies only replicated this finding, and this one-sided way of thinking appears to be a particular problem for the issues that speak to our sense of identity. Scientists today use the term "motivated reasoning" to describe this kind of emotionally charged, self-protective use of our minds. Besides the myside/confirmation bias that Perkins examined (where we preferentially seek and remember the information that confirms our view), motivated reasoning may also take the form of a *disconfirmation* bias—a kind of preferential skepticism that tears down alternative arguments. And, together, they can lead us to become more and more entrenched in our opinions.

Consider an experiment by Dan Kahan at Yale Law School, which examined attitudes to gun control. He told his participants that a local government was trying to decide whether to ban firearms in public—and it was unsure whether this would increase or decrease crime rates. So they had collected data on cities with and without these bans, and on changes in crime over one year:

	Decrease in crime	Increase in crime
Cities that banned carrying handguns in public	223	75
Cities that did not ban carrying handguns in public	107	21

Kahan also gave his participants a standard numeracy test, and questioned them on their political beliefs.

Try it for yourself. Given this data, do the bans work?

Kahan had deliberately engineered the numbers to be deceptive at first glance, suggesting a huge decrease in crime in the cities carrying the ban. To get to the correct answer, you need to consider the ratios, showing around 25 percent of the cities with the ban had witnessed an increase in crime, compared with 16 percent of those without a ban. The ban did not work, in other words.

As you might hope, the more numerate participants were more likely to come to that conclusion—but *only if they were more conservative, Republican voters who were already more likely to oppose gun control.* If they were liberal, Democratic voters, the participants skipped the explicit calculation, and were more likely to go with their (incorrect) initial hunch that the ban had worked, no matter what their intelligence.

In the name of fairness, Kahan also conducted the same experiment, but with the data reversed, so that the data supported the ban. Now, it was the numerate liberals who came to the right answer, while the numerate conservatives trusted their (incorrect) instincts. Overall, the most numerate participants were around 45 percent more likely to read the data correctly if it conformed to their expectations.

The upshot, according to Kahan and other scientists studying motivated reasoning, is that smart people do not apply their superior intelligence fairly, but instead use it "opportunistically" to promote their own interests and protect the beliefs that are most important to their identities. Intelligence can be a tool for propaganda rather than truth-seeking.[34]

It's a powerful finding, capable of explaining the enormous polarization on issues such as climate change (see graph below).[35] The scientific consensus is that carbon emissions from human sources are leading to global warming, and people with liberal politics are more likely to accept this message if they have better numeracy skills and basic scientific knowledge.[36] That makes sense, since these people should also be more likely to understand the evidence. But among free-market capitalists, the opposite is true: the more scientifically literate and numerate they are, the more likely they are to reject the scientific consensus and to believe that claims of climate change have been exaggerated.

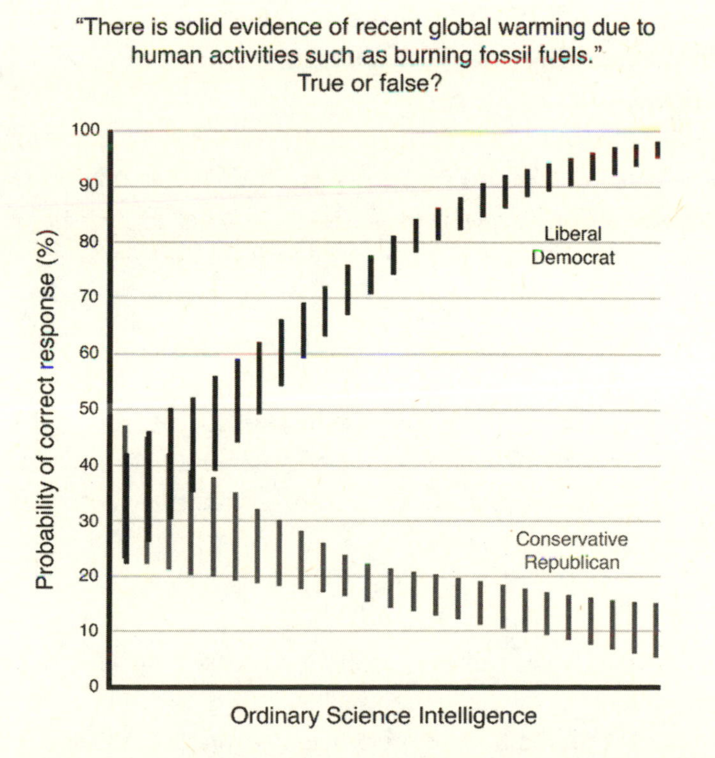

Source: Kahan, D. M. (2017). 'Ordinary science intelligence': A science-comprehension measure for study of risk and science communication, with notes on evolution and climate change. *Journal of Risk Research, 20*(8), 995–1016.

The same polarization can be seen for people's views on vaccination,[37] fracking,[38] and evolution.[39] In each case, greater education and intelligence simply helps people to justify the beliefs that match their political, social, or religious identity. (To be absolutely clear, overwhelming evidence shows that vaccines are safe and effective, carbon emissions are changing the climate, and evolution is true.)

There is even some evidence that, thanks to motivated reasoning, exposure to the opposite point of view may actually backfire; not only do people reject the counter-arguments, but their own views become even more deeply entrenched as a result. In other words, an intelligent person with an inaccurate belief system may become *more* ignorant after having heard the actual facts. We could see this with Republicans' opinions about Obamacare in 2009 and 2010: people with greater intelligence were more likely to believe claims that the new system would bring about Orwellian "death panels" to decide who lived and died, and their views were only reinforced when they were presented with evidence that was meant to debunk the myths.[40]

Kahan's research has primarily examined the role of motivated reasoning in political decision making—where there may be no right or wrong answer—but he says it may stretch to other forms of belief. He points to a study by Jonathan Koehler, then at the University of Texas at Austin, who presented parapsychologists and skeptical scientists with data on two (fictional) experiments into extrasensory perception.

The participants should have objectively measured the quality of the papers and the experimental design. But Koehler found that they often came to very different conclusions, depending on whether the results of the studies agreed or disagreed with their own beliefs in the paranormal.[41]

~

When we consider the power of motivated reasoning, Conan Doyle's belief in fraudulent mediums seems less paradoxical. His very identity had come to rest on his experiments with the paranormal. Spiritualism was the foundation of his relationship with his wife, and many of his friendships; he had invested substantial sums of money in a spiritualist church[42] and written more than twenty books and pamphlets on the subject. Approaching old age, his beliefs also provided him with the comforting certainty of the afterlife. "It absolutely removes all fear of death," he said, and the belief connected him with those he had already lost[43]—surely two of the strongest motivations imaginable.

All of this would seem to chime with research showing that beliefs may first arise from emotional needs—and it is only afterward that the intellect kicks in to rationalize the feelings, however bizarre they may be.

Conan Doyle certainly claimed to be objective. "In these 41 years, I never lost any opportunity of reading and studying and experimenting on this matter,"[44] he boasted toward the end of his life. But he was only looking for the evidence that supported his point of view, while dismissing everything else.[45]

It did not matter that this was the mind that created Sherlock Holmes—the "perfect reasoning and observing machine." Thanks to motivated reasoning, Conan Doyle could simply draw on that same creativity to explain away Houdini's skepticism. And when he saw the photos of the Cottingley Fairies, he felt he had found the proof that would convince the world of other psychic phenomena. In his excitement, his mind engineered elaborate scientific explanations—without seriously questioning whether it was just a schoolgirl joke.

When they confessed decades after Conan Doyle's death, the girls revealed that they simply hadn't bargained for grown-ups' desire to be fooled. "I never even thought of it as being a fraud," one of the girls, Frances Griffiths, revealed in a 1985 interview. "It was just Elsie and I having a bit of fun and I can't understand to this day why they were taken in—*they wanted to be taken in.*"[46]

Following their increasingly public disagreement, Houdini lost all respect for Conan Doyle; he had started the friendship believing that the writer was an "intellectual giant" and ended it by writing that "one must be half-witted to believe some of these things." But given what we know about motivated reasoning, the very opposite may be true: only an intellectual giant could be capable of believing such things.*

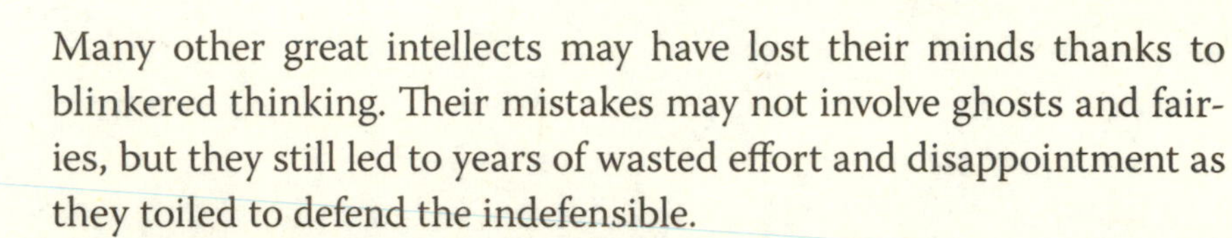

Many other great intellects may have lost their minds thanks to blinkered thinking. Their mistakes may not involve ghosts and fairies, but they still led to years of wasted effort and disappointment as they toiled to defend the indefensible.

Consider Albert Einstein, whose name has become a synonym for genius. While still working as young patent clerk in 1905, he outlined the foundations for quantum mechanics, special relativity, and the equation for mass-energy equivalence ($E=MC^2$)—the concept for which he is most famous.[47] A decade later he would announce his theory of general relativity—tearing through Isaac Newton's laws of gravity.

But his ambitions did not stop there. For the remainder of his life, he planned to build an even grander, all-encompassing understanding of the universe that melded the forces of electromagnetism and gravity into a single, unified theory. "I want to know how God created this world. I am not interested in this or that phenomenon, in the spectrum of this or that element, I want to know his thoughts," he had written previously—and this was his attempt to capture those thoughts in their entirety.

After a period of illness in 1928, he thought he had done it. "I have laid a wonderful egg. . . . Whether the bird emerging from this will be viable and long-lived lies in the lap of the gods," he wrote. But the gods soon killed that bird, and many more dashed hopes would follow over the next twenty-five years, with further announcements of a new Unified Theory, only for them all to fall like a deadweight.

---

* In his book *The Rationality Quotient*, Keith Stanovich points out that George Orwell famously came to much the same conclusion when describing various forms of nationalism, Orwell writing that: "There is no limit to the follies that can be swallowed if one is under the influence of feelings of this kind. . . . One has to belong to the intelligentsia to believe things like that: No ordinary man could be such a fool."

Soon before his death, Einstein had to admit that "most of my offspring end up very young in the graveyard of disappointed hopes."

Einstein's failures were no surprise to those around him, however. As his biographer, the physicist Hans Ohanian, wrote in his book *Einstein's Mistakes:* "Einstein's entire program was an exercise in futility. . . . It was obsolete from the start." The more he invested in the theory, however, the more reluctant he was to let it go. Freeman Dyson, a colleague at Princeton, was apparently so embarrassed by Einstein's foggy thinking that he spent eight years deliberately avoiding him on campus.

The problem was that Einstein's famous intuition—which had served him so well in 1905—had led him seriously astray, and he had become deaf and blind to anything that might disprove his theories. He ignored evidence of nuclear forces that were incompatible with his grand idea, for instance, and came to despise the results of quantum theory—a field he had once helped to establish.[48] At scientific meetings, he would spend all day trying to come up with increasingly intricate counter-examples to disprove his rivals, only to have been disproved by the evening.[49] He simply "turned his back on experiments" and tried to "rid himself of the facts," according to his colleague at Princeton, Robert Oppenheimer.[50]

Einstein himself realized as much toward the end of his life. "I must seem like an ostrich who forever buries its head in the relativistic sand in order not to face the evil quanta," he once wrote to his friend, the quantum physicist Louis de Broglie. But he continued on his fool's errand, and even on his deathbed, he scribbled pages of equations to support his erroneous theories, as the last embers of his genius faded. All of which sounds a lot like the sunk cost fallacy exacerbated by motivated reasoning.

The same stubborn approach can be found in many of his other ideas. Having supported communism, he continually turned a blind eye to the failings of the USSR, for instance.[51]

Einstein, at least, had not left his domain of expertise. But this single-minded determination to prove oneself right may be particularly damaging when scientists stray outside their usual territory, a fact that was noted by the psychologist Hans Eysenck. "Scientists, especially when they leave the particular field in which they are specialized, are just as ordinary, pig-headed, and unreasonable as everybody else," he wrote in the 1950s. "And their unusually high intelligence only makes their prejudices all the more dangerous."[52] The irony is that Eysenck himself came to believe theories of the paranormal, showing the blinkered analysis of evidence he claimed to deplore.

Some science writers have even coined a term—Nobel Disease—to describe the unfortunate habit of Nobel Prize winners to embrace dubious positions on various issues. The most notable case is, of course, Kary Mullis, the famous biochemist with the strange conspiracy theories whom we met in the introduction. His autobiography, *Dancing Naked in the Mind Field*, is almost a textbook in the contorted explanations the intelligent mind can conjure to justify its preconceptions.[53]

Other examples include Linus Pauling, who discovered the nature of chemical bonds between atoms, yet spent decades falsely claiming that vitamin supplements could cure cancer;[54] and Luc Montagnier, who helped discover the HIV virus, but who has since espoused some bizarre theories that even highly diluted DNA can cause structural changes to water, leading it to emit electromagnetic radiation. Montagnier believes that this phenomenon can be linked to autism, Alzheimer's disease, and various serious conditions, but many other scientists reject these claims, leading to a petition of 35 other Nobel laureates asking for him to be removed from his position in an AIDS research center.[55]

Although we may not be working on a Grand Unified Theory, there is a lesson here for all of us. Whatever your profession, the toxic combination of motivated reasoning and the bias blind spot could still lead us to justify prejudiced opinions about those around us, pursue failing projects at work, or rationalize a hopeless love affair.

~

As two final examples, let's look at two of history's greatest innovators: Thomas Edison and Steve Jobs.

With more than a thousand patents to his name, Thomas Edison was clearly in possession of an extraordinarily fertile mind. But once he had conceived an idea, he struggled to change his mind—as shown in the "battle of the currents."

In the late 1880s, having produced the first working electric lightbulb, Edison sought to find a way to power America's homes. His idea was to set up a power grid using a steady "direct current" (DC), but his rival George Westinghouse had found a cheaper means of transmitting electricity with the alternating current (AC) we use today. Whereas DC is a flat line of a single voltage, AC oscillates rapidly between two voltages, which stops it losing energy over distance.

Edison claimed that AC was simply too dangerous, since it more easily leads to death by electrocution. Although this concern was legitimate, the risk could be reduced with proper insulation and regulations, and the economic arguments were just too strong to ignore: it really was the only feasible way to provide electricity to the mass market.

The rational response would have been to try to capitalize on the new technology and improve its safety, rather than continuing to pursue DC. One of Edison's own engineers, Nikola Tesla, had already told him as much. But rather than taking his advice, Edison dismissed Tesla's ideas and even refused to pay him for his research into AC, leading Tesla to take his ideas to Westinghouse instead.[56]

Refusing to admit defeat, Edison engaged in an increasingly bitter PR war to try to turn public opinion against AC. It began with macabre public demonstrations, electrocuting stray dogs and horses. And when Edison heard that a New York court was investigating the possibility of using electricity for executions, he saw yet another opportunity to prove that point, as he advised the court on the development of the electric chair—in the hope that AC would be forever associated with death. It was a shocking moral sacrifice for someone who had once declared that he would "join heartily in an effort to totally abolish capital punishment."[57]

You may consider these to be simply the actions of a ruthless businessman, but the battle really was futile. As one journal stated in

1889: "It is impossible now that any man, or body of men, should resist the course of alternating current development. . . . Joshua may command the sun to stand still, but Mr. Edison is not Joshua."[58] By the 1890s, he had to admit defeat, eventually turning his attention to other projects.

The historian of science Mark Essig writes that "the question is not so much why Edison's campaign failed as why he thought it might succeed."[59] But an understanding of cognitive errors such as the sunk cost effect, the bias blind spot, and motivated reasoning helps to explain why such a brilliant mind may persuade itself to continue down such a disastrous path.

The cofounder of Apple, Steve Jobs, was similarly a man of enormous intelligence and creativity, yet he too sometimes suffered from a dangerously skewed perception of the world. According to Walter Isaacson's official biography, his acquaintances described a "reality distortion field"—"a confounding mélange of charismatic rhetorical style, indomitable will, and eagerness to bend any fact to fit the purpose at hand," in the words of his former colleague Andy Hertzfeld.

That single-minded determination helped Jobs to revolutionize technology, but it also backfired in his personal life, particularly after he was diagnosed with pancreatic cancer in 2003. Ignoring his doctor's advice, he instead opted for quack cures such as herbal remedies, spiritual healing, and a strict fruit juice diet. According to all those around him, Jobs had convinced himself that his cancer was something he could cure himself, and his amazing intelligence seems to have allowed him to dismiss any opinions to the contrary.[60]

By the time he finally underwent surgery, the cancer had progressed too far to be treatable, and some doctors believe Jobs may still have been alive today if he had simply followed medical advice. In each case, we see that greater intellect is used for rationalization and justification, rather than logic and reason.

~

We have now seen three broad reasons why an intelligent person may act stupidly. They may lack elements of creative or practical intelligence that are essential for dealing with life's challenges; they may suffer from "dysrationalia," using biased intuitive judgments to make decisions; and they may use their intelligence to dismiss any evidence that contradicts their views thanks to motivated reasoning.

Harvard University's David Perkins described this latter form of the intelligence trap to me best when he said it was like "putting a moat around a castle." The writer Michael Shermer, meanwhile, describes it as creating "logic-tight compartments" in our thinking. But I personally prefer to think of it as a runaway car, without the right steering or navigation to correct its course. As Descartes had originally put it: "those who go forward but very slowly can get further, if they always follow the right road, than those who are in too much of a hurry and stray off it."

Whatever metaphor you choose, the question of why we evolved this way is a serious puzzle for evolutionary psychologists. When they build their theories of human nature, they expect common behaviors to have had a clear benefit to our survival. But how could it ever be an advantage to be intelligent but irrational?

One compelling answer comes from the recent work of Hugo Mercier at the French National Center for Scientific Research, and Dan Sperber at the Central European University in Budapest. "I think it's now so obvious that we have the myside bias, that psychologists have forgotten how weird it is," Mercier told me in an interview. "But if you look at it from an evolutionary point of view, it's really maladaptive."

It is now widely accepted that human intelligence evolved, at least in part, to deal with the cognitive demands of managing more complex societies. Evidence comes from the archaeological record, which shows that our skull size did indeed grow as our ancestors started to live in bigger groups.[61] We need brainpower to keep track of others' feelings, to know whom you can trust, who will take advantage, and whom you need to keep sweet. And once language evolved, we needed to be eloquent, to be able to build support within the group and bring others to our way of thinking. Those arguments didn't need to be logical to bring us those benefits; they just had to be persuasive. And that subtle difference may explain why irrationality and intelligence often go hand in hand.[62]

Consider motivated reasoning and the myside bias. If human thought is primarily concerned with truth-seeking, we should weigh up both sides of an argument carefully. But if we just want to persuade others that we're right, then we're going to seem more convincing if we can pull as much evidence for our view together. Conversely, to avoid being duped ourselves, we need to be especially skeptical of others' arguments, and so we should pay extra attention to interrogating and challenging any evidence that disagrees with our own beliefs—just as Kahan had shown.

Biased reasoning isn't just an unfortunate side effect of our increased brainpower, in other words—it may have been its *raison d'être.*

In the face-to-face encounters of our ancestors' small gatherings, good arguments should have counteracted the bad, enhancing the overall problem solving to achieve a common goal; our biases could be tempered by others. But Mercier and Sperber say these mechanisms can backfire if we live in a technological and social bubble, and miss the regular argument and counterargument that could correct our biases. As a result, we simply accumulate more information to accommodate our views.

Before we learn how to protect ourselves from those errors, we must first explore one more form of the intelligence trap—"the curse of expertise," which describes the ways that acquired knowledge and professional experience (as opposed to our largely innate general intelligence) can also backfire. As we shall see in one of the FBI's most notorious mix-ups, you really can know too much.

3

The curse of knowledge: *The beauty and fragility of the expert mind*

One Friday evening in April 2004, the lawyer Brandon Mayfield made a panicked call to his mother. "If we were to somehow suddenly disappear . . . if agents of the government secretly sweep in and arrest us, I would like your assurance that you could come to Portland on the first flight and take the kids back to Kansas with you," he told her.[1]

An attorney and former officer in the US Army, Mayfield was not normally prone to paranoia, but America was still reeling from the fallout of 9/11. As a Muslim convert, married to an Egyptian wife, Mayfield sensed an atmosphere of "hysteria and islamophobia," and a series of strange events now led him to suspect that he was the target of investigation.

One day his wife Mona had returned home from work to find that the front door was double-locked with a top bolt, when the family never normally used the extra precaution. Another day, Mayfield walked into his office to find a dusty footprint on his office desk, under a loose tile on the ceiling, even though no one should have entered the room overnight. On the road, meanwhile, a mysterious car, driven by a stocky fifty- or sixty-year-old, seemed to have followed him to and from the mosque.

Given the political climate, he feared he was under surveillance. "There was this realization that it could be a secret government agency," he told me in an interview. By the time Mayfield made that impassioned phone call to his mother, he said, he had begun to feel an "impending doom" about his fate, and he was scared about what that would mean for his three children.

At around 9:45 a.m. on May 6, those fears were realized with three loud thumps on his office door. Two FBI agents had arrived to arrest Mayfield in connection with the horrendous Madrid bombings, which had killed 192 people and injured around two thousand on March 11 that year. His hands were cuffed behind his back, and he was bundled into a car and taken to the local courthouse.

He pleaded that he knew nothing of the attacks; when he first heard the news he had been shocked by the "senseless violence," he said. But FBI agents claimed to have found his fingerprint on a blue shopping bag containing detonators, left in a van in Madrid. The FBI declared it was a "100% positive match"; there was no chance they were wrong.

As he describes in his book, *Improbable Cause*, Mayfield was held in a cell while the FBI put together a case to present to the Grand Jury. He was handcuffed and shackled in leg irons and belly chains, and subjected to frequent strip searches.

His lawyers painted a bleak picture: if the Grand Jury decided he was involved in the attacks, he could be shipped to Guantanamo Bay. As the judge stated in his first hearing, fingerprints are considered the gold standard of forensic evidence: people had previously been convicted for murder based on little more than a single print. The chances of two people sharing the same fingerprint were considered to be billions to one.[2]

Mayfield tried to conceive how his fingerprint could have appeared on a plastic carrier bag more than 5,400 miles away—across the entire American continent and Atlantic Ocean. But there was no way. His lawyers warned that the very act of denying such a strong line of evidence could mean that he was indicted for perjury. "I thought I was being framed by unnamed officials—that was the immediate thought," Mayfield told me.

His lawyers eventually persuaded the court to employ an independent examiner, Kenneth Moses, to reanalyze the prints. Like those of the FBI's own experts, Moses's credentials were impeccable. He had served with the San Francisco Police Department for twenty-seven years, and had garnered many awards and honors during his service.[3] It was Mayfield's last chance, and on May 19—after nearly two weeks in prison—he returned to the tenth floor of the courthouse, to hear Moses give his testimony by video conference.

As Moses's testimony unfurled, Mayfield's worst fears were confirmed. "I compared the latent prints to the known prints that were submitted on Brandon Mayfield," Moses told the court. "And I concluded that the latent print is the left index finger of Mr Mayfield."[4]

Little did he know that a remarkable turn of events taking place on the other side of the Atlantic Ocean would soon save him. That very morning, the Spanish National Police had identified an Algerian man, Ouhnane Daoud, connected with the bombings. Not only could they show that his finger better fitted the print previously matched to Mayfield—including some ambiguous areas dismissed by the FBI—but his thumb also matched an additional print found on the bag. He was definitely their man.

Mayfield was freed the next day, and by the end of the month, the FBI would have to release a humiliating public apology.

What went wrong? Of all the potential explanations, a simple lack of skill cannot be the answer: the FBI's forensics teams are considered to be the best in the world.[5] Indeed, a closer look at the FBI's mistakes reveals that they did not occur *despite* its examiners' knowledge—they may have occurred *because of it.*

~

The previous chapters have examined how general intelligence—the capacity for abstract reasoning measured by IQ or SATs—can backfire. The emphasis here should be on the word *general*, though, and you might hope we could mitigate those errors through more specialized knowledge and professional expertise, cultivated through years of experience. Unfortunately, the latest research shows that these can also lead us to err in unexpected ways.

These discoveries should not be confused with some of the vaguer criticisms that academics (such as Paul Frampton) live in an "ivory tower" isolated from "real life." Instead, the latest research highlights dangers in the exact situations where most people would hope that experience protects you from mistakes.

If you are undergoing heart surgery, flying across the globe, or looking to invest an unexpected windfall, you want to be in the care of a doctor, pilot, or accountant with a long and successful career behind them. If you want an independent witness to verify a fingerprint match in a high-profile case, you choose Moses. Yet there are now various social, psychological, and neurological reasons that explain why expert judgment sometimes fails at the times when it is most needed—and the sources of these errors are intimately entwined with the very processes that normally allow experts to perform so well.

"A lot of the cornerstones, the building blocks that make the expert an expert and allow them to do their job efficiently and quickly, also entail vulnerabilities: you can't have one without the other," explains the cognitive neuroscientist Itiel Dror at University College London, who has been at the forefront of much of this research. "The more expert you are, the more vulnerable you are in many ways."

Clearly experts will still be right the majority of times, but when they are wrong, it can be disastrous, and a clear understanding of the overlooked potential for expert error is essential if we are to prevent those failings.

As we shall soon discover, those frailties blinded the FBI examiners' judgment—bringing about the string of bad decisions that led to Mayfield's arrest. In aviation they have led to the unnecessary deaths of pilots and civilians, and in finance they contributed to the 2008 financial crisis.

~

Before we examine that research, we first need to consider some core assumptions. One potential source of expert error could be a sense of overconfidence. Perhaps experts overreach themselves, believing their powers are infallible? The idea would seem to fit with the descriptions of the bias blind spot that we explored in the last chapter.

Until recently, however, the bulk of the scientific research suggested the opposite was true: it's the incompetents who have an inflated view of their abilities. Consider a classic study by David Dunning at the University of Michigan and Justin Kruger at New York University. Dunning and Kruger were apparently inspired by the unfortunate case of McArthur Wheeler, who attempted to rob two banks in Pittsburgh in 1995. He committed the crimes in broad daylight, and the police arrested him just hours later. Wheeler was genuinely perplexed. "But I wore the juice!" he apparently exclaimed. Wheeler, it turned out, believed a coating of lemon juice (the basis of invisible ink) would make him imperceptible on the CCTV footage.[6]

From this story, Dunning and Kruger wondered if ignorance often comes hand in hand with overconfidence, and set about testing the idea in a series of experiments. They gave students tests on grammar and logical reasoning, and then asked them to rate how well they thought they had performed. Most people misjudged their own abilities, but this was particularly true for the people who performed the most poorly. In technical terms, their confidence was poorly calibrated—they simply had no idea just how bad they were. Crucially, Dunning and Kruger found that they could reduce that overconfidence by offering training in the relevant skills. Not only did the participants get better at what they did; their increased knowledge also helped them to understand their limitations.[7]

Since Dunning and Kruger first published their study in 1999, the finding has been replicated many times, across many different cultures.[8] One survey of thirty-four countries—from Australia to Germany, and Brazil to South Korea—examined the maths skills of fifteen-year-old students; once again, the least able were often the most overconfident.[9]

Unsurprisingly, the press have been quick to embrace the "Dunning-Kruger Effect," declaring that it is the reason why "losers have delusions of grandeur" and "why incompetents think they are awesome" and citing it as the cause of President Donald Trump's more egotistical statements.[10]

The Dunning-Kruger Effect should have an upside, though. Although it may be alarming when someone who is highly incompetent but confident reaches a position of power, it does at least reassure us that education and training work as we would hope, improving not just our knowledge but our metacognition and self-awareness. This was, incidentally, Bertrand Russell's thinking in an essay called "The Triumph of Stupidity" in which he declared that "the fundamental cause of the trouble is that in the modern world the stupid are cocksure while the intelligent are full of doubt."

Unfortunately, these discoveries do not paint the whole picture. In charting the shaky relationship between perceived and actual competence, these experiments had focused on general skills and knowledge, rather than the more formal and extensive study that comes with a university degree, for example.[11] And when you do investigate people with an advanced education, a more unsettling vision of the expert brain begins to emerge.

In 2010, a group of mathematicians, historians and athletes were tasked with identifying certain names that represented significant figures within each discipline. They had to discern whether Johannes de Groot or Benoit Theron were famous mathematicians, for instance, and they could answer, Yes, No, or Don't Know. As you might hope, the experts were better at picking out the right people (such as Johannes de Groot, who really was a mathematician) if they fell within their discipline. But they were also more likely to say they recognized the made-up figures (in this case, Benoit Theron).[12] When their self-perception of expertise was under question, they would rather take a guess and "over-claim" the extent of their knowledge than admit their ignorance with a "don't know."

Matthew Fisher at Yale University, meanwhile, quizzed university graduates on their college major for a study published in 2016. He wanted to check their knowledge of the core topics of the degree, so he first asked them to estimate how well they understood some of the fundamental principles of their discipline; a physicist might have been asked to gauge their understanding of thermodynamics; a biologist, to describe Kreb's Cycle.

Unbeknown to the participants, Fisher then sprung a surprise test: they now had to write a detailed description of the principles they claimed to know. Despite having declared a high level of knowledge, many stumbled and struggled to write a coherent explanation. Crucially, this was only true within the topic of their degree. When graduates also considered topics beyond their specialization, or more general, everyday subjects, their initial estimates tended to be far more realistic.[13]

One likely reason is that the participants simply had not realized how much they might have forgotten since their degree (a phenomenon that Fisher calls meta-forgetfulness). "People confuse their current level of understanding with their peak level of knowledge," Fisher told me. And that may suggest a serious problem with our education. "The most cynical reading of it is that we're not giving students knowledge that stays with them," Fisher said. "We're just giving them the sense they know things, when they actually don't. And that seems to be counterproductive."

The illusion of expertise may also make you more closed-minded. Victor Ottati at Loyola University in Chicago has shown that priming people to feel knowledgeable means that they were less likely to seek or listen to the views of people who disagreed with them.* Ottati notes that this makes sense when you consider the social norms surrounding expertise; we assume that an expert already has the credentials to stick to their opinions, which he calls "earned dogmatism."[14]

In many cases, of course, experts really may have better justifications to think what they do. But if they overestimate their own knowledge—as Fisher's work might suggest—and then stubbornly refuse to seek or accept another's opinion, they may quickly find themselves out of their depth.

Ottati speculates that this fact could explain why some politicians become more entrenched in their opinions and fail to update their

* The Japanese, incidentally, have encoded these ideas in the word *shoshin*, which encapsulates the fertility of the beginner's mind and its readiness to accept new ideas. As the Zen monk Shunryu Suzuki put it in the 1970s: "In the beginner's mind there are many possibilities; in the expert's, there are few."

knowledge or seek compromise—a state of mind he describes as "myopic over-self-confidence."

Earned dogmatism might also further explain the bizarre claims of the scientists with "Nobel Disease" such as Kary Mullis. Subrahmanyan Chandrasekhar, the Nobel Prize–winning Indian-American astrophysicist, observed this tendency in his colleagues. "These people have had great insights and made profound discoveries. They imagine afterward that the fact that they succeeded so triumphantly in one area means they have a special way of looking at science that must be right. But science doesn't permit that. Nature has shown over and over again that the kinds of truth which underlie nature transcend the most powerful minds."[15]

~

Inflated self-confidence and earned dogmatism are just the start of the expert's flaws, and to understand the FBI's errors, we have to delve deeper into the neuroscience of expertise and the ways that extensive training can permanently change our brain's perception—for good and bad.

The story begins with a Dutch psychologist named Adriaan de Groot, who is sometimes considered *the* pioneer of cognitive psychology. Beginning his career during the Second World War, de Groot had been something of a prodigious talent at school and university—showing promise in music, mathematics, and psychology—but the tense political situation on the eve of the war offered few opportunities to pursue academia after graduation. Instead, de Groot found himself scraping together a living as a high-school teacher, and later, as an occupational psychologist for a railway company.[16]

De Groot's real passion was chess, however. A considerably talented player, he had represented his country at an international tournament in Buenos Aires,[17] and decided to interview other players about their strategies to see if they could reveal the secrets of exceptional performance.[18] He began by showing them a sample chess board before asking them to talk through their mental strategies as they decided on the next move.

De Groot had initially suspected that their talents might arise from the brute force of their mental calculations: perhaps they were simply better at crunching the possible moves and simulating the consequences. This didn't seem to be the case, however: the experts didn't report having cycled through many positions, and they often made up their minds within a few seconds, which would not have given them enough time to consider the different strategies.

Follow-up experiments revealed that the players' apparent intuition was in fact an astonishing feat of memory, achieved through a process that is now known as "chunking." The expert player stops seeing the game in terms of individual pieces and instead breaks the board into bigger units—or "complexes"—of pieces. In the same ways that words can be combined into larger sentences, those complexes can then form templates or psychological scripts known as "schemas," each of which represents a different situation and strategy. This transforms the board into something *meaningful*, and it is thought to be the reason that some chess grandmasters can play multiple games simultaneously—even while blindfolded. The use of schemas significantly reduces the processing workload for the player's brain; rather than computing each potential move from scratch, experts search through a vast mental library of schemas to find the move that fits the board in front of them.

De Groot noted that over time the schemas can become deeply "engrained in the player," meaning that the right solution may come to mind automatically with just a mere glance at the board, which neatly accounts for those phenomenal flashes of brilliance that we have come to associate with expert intuition. Automatic, engrained behaviors also free up more of the brain's working memory, which might explain how experts operate in challenging environments. "If this were not the case," de Groot later wrote, "it would be completely impossible to explain why some chess players can still play brilliantly while under the influence of alcohol."[19]

De Groot's findings would eventually offer a way out of his tedious jobs at the high school and railway, earning him a doctorate from the University of Amsterdam. And it has since inspired countless other studies in many domains—explaining the talent of everyone from Scrabble and poker champions to the astonishing performances of elite athletes like Serena Williams, and the rapid coding of world-class computer programmers.[20]

Although the exact processes will differ depending on the particular skill, in each case the expert is benefiting from a vast library of schemas that allows them to extract the most important information, recognize the underlying patterns and dynamics, and react with an almost automatic response from a pre-learned script.[21]

This theory of expertise may also help us to understand less celebrated talents, such as the extraordinary navigation of London taxi drivers through the city's 25,000 streets. Rather than remembering the whole cityscape, they have built schemas of known routes, so that the sight of a landmark will immediately suggest the best path from A to B, depending on the traffic at hand—without their having to recall and process the entire map.[22]

Even burglars may operate using the same neural processes. Asking real convicts to take part in virtual reality simulations of their crimes, researchers have demonstrated that more experienced burglars have amassed a set of advanced schemas based on the familiar layouts of British homes, allowing them to automatically intuit the best route through a house and to alight on the most valuable possessions.[23] As one prison inmate told the researchers: "The search becomes a natural instinct, like a military operation—it becomes routine."[24]

There is no denying that the expert's intuition is a highly efficient way of working in the vast majority of situations they face—and it is often celebrated as a form of almost superhuman genius.

Unfortunately, it can also come with costly sacrifices.

One is flexibility: the expert may lean so heavily on existing behavioral schemas that they struggle to cope with change.[25] When tested on their memories, experienced London taxi drivers appeared to struggle with the rapid development of Canary Wharf at the end of the twentieth century, for instance; they just couldn't integrate the new landmarks and update their old mental templates of the city.[26] Similarly, an expert games champion will find it harder to learn a new set of rules and an accountant will struggle to adapt to new tax laws. The same cognitive entrenchment can also limit creative problem solving if the expert fails to look beyond their existing schemas for new ways to tackle a challenge. They become entrenched in the familiar ways of doing things.

The second sacrifice may be an eye for detail. As the expert brain chunks up the raw information into more meaningful components, and works at recognizing broad underlying patterns, it loses sight of the smaller elements. This change has been recorded in real-time scans of expert radiologists brains: they tend to show greater activity in the areas of the temporal lobe associated with advanced pattern recognition and symbolic meaning, but less activity in regions of the visual cortex that are associated with combing over fine detail.[27] The advantage will be the ability to filter out irrelevant information and reduce distraction, but this also means the expert is less likely to consider all the elements of a problem systematically, potentially causing them to miss important nuances that do not easily fit their mental maps.

It gets worse. Expert decisions, based on gist rather than careful analysis, are also more easily swayed by emotions and expectations and cognitive biases such as framing and anchoring.[28] The upshot is that training may have actually *reduced* their rationality quotient. "The expert's mind-set—based on what they expect, what they hope, whether they are in a good mood or bad mood that day—affects how they look at the information," Itiel Dror told me. "And the brain mechanisms—the actual cognitive architecture—that give an expert their expertise are especially vulnerable to that."

The expert could, of course, override their intuitions and return to a more detailed, systematic analysis. But often they are completely unaware of the danger—they have the bias blind spot that we observed in Chapter 2.[29] The result is a kind of ceiling to their accuracy, as these errors become more common among experts than those arising from ignorance or inexperience. When that fallible, gist-based processing is combined with overconfidence and "earned dogmatism," it gives us one final form of the intelligence trap—and the consequences can be truly devastating.

~

The FBI's handling of the Madrid bombings offers the perfect example of these processes in action. Matching fingerprints is an extraordinarily complex job, with analyses based on three levels of increasingly intricate features, from broad patterns, such as whether your prints have a left- or right-facing swirl, a whorl, or an arch, to the finer details of the ridges in your skin—whether a particular line splits in two, breaks into fragments, forms a loop called an "eye," or ends abruptly. Overall, examiners may aim to detect around ten identifying features.

Eye-tracking studies reveal that expert examiners often go through this process semi-automatically,[30] chunking the picture in much the same way as de Groot's chess grandmasters[31] to identify the features that are considered the most useful for comparison. As a result, the points of identification may just jump out at the expert analyst, while a novice would have to systematically identify and check each one—making it exactly the kind of top-down decision making that can be swayed by bias.

Sure enough, Dror has found that expert examiners are prone to a range of cognitive errors that may arise from such automatic processing. They were more likely to find a positive match if they were told a suspect had already confessed to the crime.[32] The same was true when they were presented with emotive material, such as a gory picture of a murder victim. Although it should have had no bearing on their objective judgment, the examiners were again more likely to link the fingerprints, perhaps because they felt more motivated and determined to catch the culprit.[33] Dror points out that this is a particular problem when the available data is ambiguous and messy—and that was exactly the problem with the evidence from Madrid. The fingerprint had been left on a crumpled carrier bag; it was smeared and initially difficult to read.

The FBI had first run the fingerprint through a computer analysis to find potential suspects among their millions of recorded prints, and Mayfield's name appeared as the fourth of twenty possible suspects. At this stage, the FBI analysts apparently had no idea of his background—his print was only on file from a teenage brush with the law. But it seems likely that they were hungry for a match, and once they settled on Mayfield they became more and more invested in their choice—despite serious signs that they had made the wrong decision.

While the examiners had indeed identified around fifteen points of similarity in the fingerprints, they had consistently ignored significant differences. Most spectacularly, a whole section of the latent print—the upper left-hand portion—failed to match Mayfield's index finger. The examiners had argued that this area might have come from someone else's finger, who had touched the bag at another time; or maybe it came from Mayfield himself, leaving another print super-imposed on the first one to create a confusing pattern. Either way, they decided they could exclude that anomalous section and simply focus on the bit that looked most like Mayfield's.

If the anomalous section had come from another finger, however, you would expect to see tell-tale signs. The two fingers would have been at different angles, for instance, meaning that the ridges would be overlapping and criss-crossing. You might also expect that the two fingers would have touched the bag with varying pressure, affecting the appearance of the impressions left behind; one section might have seemed fainter than the first. Neither sign was present in this case.

For the FBI's story to make sense, the two people would have gripped the bag with exactly the same force, and their prints would have had to miraculously align. The chances of that happening were tiny. The much likelier explanation was that the print came from a single finger—and that finger was not Mayfield's.

These were not small subtleties but glaring holes in the argument. A subsequent report by the Office of the Inspector General (OIG) found that the complete neglect of this possibility was completely unwarranted. "The explanation required the examiners to accept an extraordinary set of coincidences," the OIG concluded.[34] Given those discrepancies, some independent fingerprint examiners reviewing the case concluded that Mayfield should have been ruled out right away.[35]

Nor was this the only example of such circular reasoning in the FBI's case: the OIG found that across the whole of their analysis, the examiners appeared far more likely to dismiss or ignore any points of interest that disagreed with their initial hunch, while showing far less scrutiny for details that appeared to suggest a match.

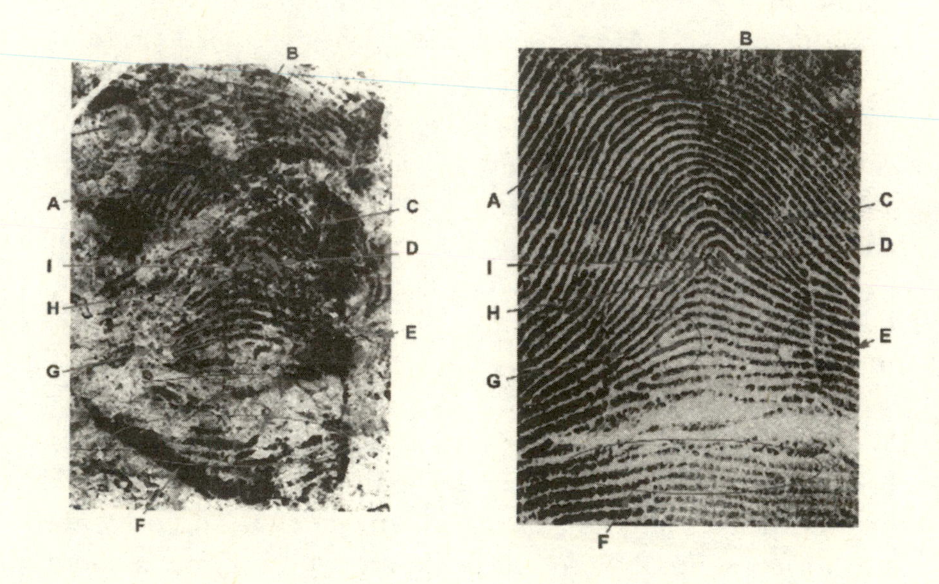

The two marked-up prints above, taken from the freely available OIG report, show just how many errors they made. The Madrid print is on the left; Mayfield's is on the right. Admittedly the errors are hard to see for a complete novice, but if you look very carefully you can make out some notable features that are present in one but not the other.

The OIG concluded that this was a clear case of the confirmation bias, but given what we have learned from the research on top-down processing and the selective attention that comes with expertise, it is possible that the examiners weren't even seeing those details in the first place. They were almost literally blinded by their expectations.

These failings could have been uncovered with a truly independent analysis. But although the prints moved through multiple examiners, each one knew their colleague's conclusions, swaying their judgment. (Dror calls this a "bias cascade.")[36] This also spread to the officers performing the covert surveillance of Mayfield and his family, who even mistook his daughter's Spanish homework for travel documents placing him in Madrid at the time of the attack.

Those biases will only have been strengthened once the FBI looked into Mayfield's past and discovered that he was a practicing Muslim, and that he had once represented one of the Portland Seven terrorists in a child custody case. In reality, it had no bearing on his presumed guilt.[37]

The FBI's confidence was so great that they ignored additional evidence from Spain's National Police (the SNP). By mid-April the SNP had tried and failed to verify the match, yet the FBI lab quickly disregarded their concerns. "They had a justification for everything," Pedro Luis Mélida Lledó, head of the fingerprint unit for the SNP, told the *New York Times* shortly after Mayfield was exonerated.[38] "But I just couldn't see it."

Records of the FBI's internal emails confirm that the examiners were unshaken by the disagreement. "I spoke with the lab this morning and they are absolutely confident that they have the match to the print—No doubt about it!!!!!" one FBI agent wrote. "They will testify in any court you swear them into."[39]

That complete conviction may have landed Mayfield in Guantanamo Bay—or death row—if the SNP had not succeeded in finding their own evidence that he was innocent. A few weeks after the original bombings, they raided a house in suburban Madrid. The suspects detonated a suicide bomb rather than submitting to arrest, but the police managed to uncover documents bearing the name of Ouhnane Daoud: an Algerian national, whose prints had been on record for an immigration event. Mayfield was released, and within a week, he was completely exonerated of any connection to the attack. Challenging the lawfulness of his arrest, he eventually received $2 million in compensation.

The lesson here is not just psychological, but social. Mayfield's case perfectly illustrates the ways that the overconfidence of experts themselves, combined with our blind faith in their talents, can amplify their biases—with potentially devastating effect. The chain of failures within the FBI and the courtroom should not have been able to escalate so rapidly, given the lack of evidence that Mayfield had even left the country.

~

With this knowledge in mind, we can begin to understand why some existing safety procedures—although often highly effective—nevertheless fail to protect us from expert error.

Consider aviation. Commonly considered to be one of the most reliable industries on Earth, airports and pilots already make use of numerous safety nets to catch any momentary lapses of judgment. The use of checklists as reminders of critical procedures—now common in many other sectors—originated in the cockpit to ensure, for instance, safer takeoffs and landings.

Yet these strategies do not account for the blind spots that specifically arise from expertise. With experience, the safety procedures are simply integrated into the pilot's automatic scripts and shrink from conscious awareness. The result, according to one study of nineteen serious accidents, is "an insidious move toward less conservative judgment" and it has led to people dying when the pilot's knowledge should have protected them from error.[40]

This was evident at Blue Grass Airport in Lexington, Kentucky, on August 25, 2007, at 6 a.m. in the morning. Comair Flight 5191 had been due to take off from runway 22 around 6 a.m., but the pilot lined up on a shorter runway. Thanks to the biases that came with their extensive experience, both the pilot and copilot missed all the warning signs that they were in the wrong place. The plane smashed through the perimeter fence, before ricocheting off an embankment, crashing into a pair of trees, and bursting into flames. Forty-seven passengers—and the pilot—died as a result.[41]

The curse of expertise in aviation doesn't end there. As we saw with the FBI's forensic scientists, experimental studies have shown that a pilot's expertise may even influence their visual perception—causing them to underestimate the depth of cloud in a storm, for instance, based on their prior expectations.[42]

The intelligence trap shows us that it's not good enough to be *fool proof*; procedures need to be *expert proof* too. The nuclear power industry is one of the few sectors to account for the automatization that comes with experience, with some plants routinely switching the order of procedures in their safety checks to prevent inspectors from working on autopilot. Many other industries, including aviation, could learn the same lesson.[43]

~

A greater appreciation of the curse of expertise—and the virtues of ignorance—can also explain how some organizations weather chaos and uncertainty, while others crumble in the changing wind.

Consider a study by Rohan Williamson of Georgetown University, who recently examined the fortunes of banks during financial crises. He was interested in the roles of "independent directors"—people recruited from outside the organization to advise the management. The independent director is meant to offer a form a self-regulation, which should require a certain level of expertise, and many do indeed come from other financial institutions. Due to the difficulties of recruiting a qualified expert without any other conflicting interests, however, some of the independent directors may be drawn from other areas of business, meaning they may lack the more technical knowledge of the processes involved in the bank's complex transactions.

Bodies such as the Organisation for Economic Cooperation and Development (OECD) had previously argued that this lack of financial expertise may have contributed to the 2008 financial crisis.[44]

But what if they'd got it the wrong way around, and this ignorance was actually a virtue? To find out, Williamson examined the data of 100 banks before and after the crisis. Until 2006, the results were exactly as you might expect if you assume that greater knowledge always aids decision making: banks with an expert board performed slightly better than those with fewer (or no) independent directors holding a background in finance since they were more likely to endorse risky strategies that paid off.

Their fortunes took a dramatic turn after the financial crash, however; now it was the banks with the least expertise that performed better. The "expert" board members, so deeply embedded in their already risky decision making, didn't pull back and adapt their strategy, while the less knowledgeable independent directors were less entrenched and biased, allowing them to reduce the banks' losses as they guided them through the crash.[45]

Although this evidence comes from finance—an area not always respected for its rationality—the lessons could be equally valuable for any area of business. When the going gets tough, the less experienced members of your team may well be the best equipped to guide you out of the mess.

~

In forensic science, at least, there has been some movement to mitigate the expert errors behind the FBI's investigations into Brandon Mayfield.

"Before Brandon Mayfield, the fingerprint community really liked to explain any errors in the language of incompetence," says the UCLA law professor Jennifer Mnookin. "Brandon Mayfield opened up a space for talking about the possibility that really good analysts, using their methods correctly, could make a mistake."[46]

Itiel Dror has been at the forefront of the work detailing these potential errors in forensic judgments, and recommending possible measures that could mitigate the effects. For example, he advocates more advanced training that includes a cognitively informed discussion of bias, so that every forensic scientist is aware of the ways their judgment may be swayed, and practical ways to minimize these influences. "Like an alcoholic at an AA meeting, acknowledging the problem is the first step in the solution," he told me.

Another requirement is that forensic analysts make their judgments "blind," without any information beyond the direct evidence at hand, so that they are not influenced by expectation but see the evidence as objectively as possible. This is especially crucial when seeking a second opinion: the second examiner should have no knowledge of the first judgment.

The evidence itself must be presented in the right way and in the right sequence, using a process that Itiel Dror calls "Linear Sequential Unmasking" to avoid the circular reasoning that had afflicted the examiners' judgment after the Madrid bombings.[47] For instance, the examiners should first mark up the latent print left on the scene before even seeing the suspect's fingerprint, giving them predetermined points of comparison. And they should not receive any information about the context of a case before making their forensic judgment of the evidence. This system is now used by the FBI and other agencies and police departments across the United States and other countries.

Dror's message was not initially welcomed by the experts he has studied; during our conversation at London's Wellcome Collection, he showed me an angry letter, published in a forensics journal, from the Chair of the Fingerprint Society, which showed how incensed many examiners were at the very idea that they may be influenced by their expectations and their emotions. "Any fingerprint examiner who comes to a decision on identification and is swayed either way in that decision-making process under the influence of stories and gory images is either totally incapable of performing the noble tasks expected of him/her or is so immature he/she should seek employment at Disneyland," he wrote.

Recently, however, Dror has found that more and more forces are taking his suggestions on board. "Things are changing . . . but it's slow. You will still find that if you talk to certain examiners, they will say 'Oh no, we're objective.'"

Mayfield retains some doubts about whether these were genuine unconscious errors, or the result of a deliberate set-up, but he supports any work that helps to highlight the frailties of fingerprint analysis. "In court, each piece of evidence is like a brick in a wall," he told me. "The problem is that they treat the fingerprint analysis as if it is the whole wall—but it's not even a strong brick, let alone a wall."

Mayfield continues to work as a lawyer. He is also an active campaigner, and has cowritten his account of the ordeal, called *Improbable Cause*, with his daughter Sharia, in a bid to raise awareness of the erosion of US civil liberties in the face of more stringent government surveillance. During our conversation, he appeared to be remarkably stoic about his ordeal. "I'm talking to you—I'm not locked in Guantanamo, in some Kafkaesque situation. . . . So in that sense, the justice system must have worked," he told me. "But there may be many more people who are not in such an enviable position."

~

With this knowledge in mind, we are now ready to start Part 2. Through the stories of the Termites, Arthur Conan Doyle, and the FBI's forensic examiners, we have seen four potential forms of the intelligence trap:

- We may lack the necessary *tacit knowledge* and *counterfactual thinking* that are essential for executing a plan and preempting the consequences of your actions.
- We may suffer from *dysrationalia, motivated reasoning,* and the *bias blind spot,* which allow us to rationalize and perpetuate our mistakes, without recognizing the flaws in our own thinking. This results in our building "logic-tight compartments" around our beliefs without considering all the available evidence.
- We may place too much confidence in our own judgment, thanks to *earned dogmatism,* so that we no longer perceive our limitations and overreach our abilities.
- Finally, thanks to our expertise, we may employ *entrenched, automatic behaviors* that render us oblivious to the obvious warning signs that disaster is looming, and more susceptible to bias.

If we return to the analogy of the brain as a car, this research confirms the idea that intelligence is the engine, and education and expertise are its fuel; by equipping us with the basic abstract reasoning skills and specialist knowledge, they put our thinking in motion, but simply adding more power won't always help you to drive that vehicle safely. Without counterfactual thinking and tacit knowledge, you may find yourself up a dead end; if you suffer from motivated reasoning, earned dogmatism, and entrenchment, you risk simply driving in circles, or worse, off a cliff.

Clearly, we've identified the problem, but we are still in need of some lessons to teach us how to navigate these potential pitfalls more carefully. Correcting these omissions is now the purpose of a whole new scientific discipline—evidence-based wisdom—which we shall explore in Part 2.

PART 2

Escaping the intelligence trap:
A toolkit for reasoning and decision making

4

Moral algebra: *Toward the science of evidence-based wisdom*

We are in the stuffy State House of Pennsylvania, in the summer of 1787. It is the middle of a stifling heat wave, but the windows and doors have been locked against the prying eyes of the public, and the sweating delegates—many dressed in thick woollen suits[1]—are arguing fiercely. Their aim is to write the new US Constitution—and the stakes could not be higher. Just eleven years after the American colonies declared independence from England, the country's government is underfunded and nearly impotent, with serious infighting between the states. It's clear that a new power structure is desperately needed to pull the country together.

Perhaps the thorniest issue concerns how the public will be represented in Congress. Would the representatives be chosen by a popular vote, or selected by local governments? Should larger states have more seats? Or should each state be given equal representation—regardless of its size? Smaller states such as Delaware fear they could be dominated by larger states such as Virginia.[2]

With tempers as hot as the sweltering weather, the closed State House proves to be the perfect pressure cooker, and by the end of the summer the Convention looks set to self-combust. It falls to Benjamin Franklin—Philadelphia's own delegate—to relieve the tension.

At eighty-one, Franklin is the oldest man at the Convention, and the once robust and hearty man is now so frail that he is sometimes carried into the proceedings on a sedan chair. Having personally signed the Declaration of Independence, he fears that America's reputation in the eyes of the world hinges on their success. "If it does not do good it will do harm, as it will show that we have not the wisdom enough among us to govern ourselves," he had previously written to Thomas Jefferson, who was abroad at the time.[3]

Franklin plays the role of the pragmatic host: after the day's debating is over, he invites the delegates to eat and drink in his garden, just a few hundred feet from the Convention, where he may encourage calmer discussion under the cooling shade of his mulberry tree. He sometimes brings out his scientific collection, including a prized two-headed snake—which he uses as a metaphor for indecision and disagreement.

In the State House itself, Franklin is often silent, and largely influences the discussions through pre-written speeches. But when he does intervene, he pleads for compromise. "When a broad table is to be made, and the edges of planks do not fit, the artist takes a little from both, and makes a good joint," he argues during one heated debate in June.[4]

This pragmatic "carpentry" eventually presents a solution for the issue of states' representation—a problem that was fast threatening to destroy the Convention. The idea came from Roger Sherman and Oliver Ellsworth, two delegates from Connecticut, who proposed that Congress could be divided into two houses, each voted for with a different system. In the Lower House, representatives would be apportioned according to population size (pleasing the larger states) while the Senate would have an equal number of delegates per state, regardless of size (pleasing the smaller states).

The "Great Compromise" is at first rejected by the delegates—until Franklin becomes its champion. He refines the proposal—arguing that the House would be in charge of taxation and spending; the Senate would deal with matters of state sovereignty and executive orders—and it is finally approved in a round of voting.

On September 17, it is time for the delegates to decide whether to put their names to the finished document. Even now, success is not inevitable until Franklin closes the proceedings with a rousing speech.

"I confess that there are several parts of this constitution which I do not at present approve, but I am not sure I shall never approve them," he declares.[5] "For having lived long, I have experienced many instances of being obliged by better information, or fuller consideration, to change opinions even on important subjects, which I once thought right, but found to be otherwise. It is therefore that the older I grow, the more apt I am to doubt my own judgment, and to pay more respect to the judgment of others."

It is only right, he says, that a group of such intelligent and diverse men should come along with their own prejudices and passions—but he ends by asking them to consider that their judgments might be wrong. "I cannot help expressing a wish that every member of the Convention who may still have objections to it, would with me, on this occasion, doubt a little of his own infallibility, and to make manifest our unanimity, put his name to this instrument."

The delegates take his advice and, one by one, the majority sign the document. Relieved, Franklin looks to George Washington's chair, with its engraving of the sun on the horizon. He has long pondered the direction of its movement. "But now at length I have the happiness to know that it is a rising and not a setting sun."

Franklin's calm, stately reasoning is a stark contrast to the biased, myopic thinking that so often comes with great intelligence and expertise. He was, according to his biographer Walter Isaacson, "allergic to anything smacking of dogma." He combined this open-minded attitude with practical good sense, incisive social skills, and astute emotional regulation—"an empirical temperament that was generally averse to sweeping passions."[6]

He wasn't always enlightened on every issue. His early views on slavery, for instance, are indefensible, although he later came to be the president of the Pennsylvania Abolition Society. But in general—and particularly in later life—he managed to navigate extraordinarily complex dilemmas with astonishing wisdom.

This same mind-set had already allowed him to negotiate an alliance with France, and a peace treaty with Britain, during the War of Independence, leading him to be considered, according to one scholar, "the most essential and successful American diplomat of all time."[7] And at the signing of the Constitution, it allowed him to guide the delegates to the solution of an infinitely complex and seemingly intractable political disagreement.

Fortunately, psychologists are now beginning to study this kind of mind-set in the new science of "evidence-based wisdom." Providing a direct contrast to our previously narrow understanding of human reasoning, this research gives us a unifying theory that explains many of the difficulties we have explored so far, while also providing practical techniques to cultivate wiser thinking and escape the intelligence trap.

As we shall see, the same principles can help us think more clearly about everything from our most personal decisions to important world events; the same strategies may even lie behind the astonishing predictions of the world's "super-forecasters."

~

First, some definitions. In place of esoteric or spiritual concepts of wisdom, this scientific research has focused on secular definitions, drawn from philosophy, including Aristotle's view of practical wisdom—"the set of skills, dispositions, and policies that help us understand and deliberate about what's good in life and helps us to choose the best means for pursuing those things over the course of the life," according to the philosopher Valerie Tiberius. (This was, incidentally, much the same definition that Franklin used.)[8] Inevitably, those skills and characteristics could include elements of the "tacit knowledge" we explored in Chapter 1, and various social and emotional skills, as well as encompassing the new research on rationality. "Now if you want to be wise it's important to know we have biases like that and it's important to know what policies you could enact to get past those biases," Tiberius said.[9]

Even so, it is only relatively recently that scientists have devoted themselves to the study of wisdom as its own construct.[10] The first steps toward a more empirical framework came in the 1970s, with ethnographic research exploring how people experience wisdom in their everyday lives, and questionnaires examining how elements of thinking associated with wisdom—such as our ability to balance different interests—change over a lifetime. Sure enough, wise reasoning did seem to increase with age.

Robert Sternberg (who had also built the scientific definitions of practical and creative intelligence that we explored in Chapter 1) was a prominent champion of this early work and helped to cement its credibility; the work has even inspired some of the questions in his university admission tests.[11]

An interest in a scientifically well-defined measure of wisdom would only grow following the 2008 financial crash. "There was a kind of social disapprobation for 'cleverness' at the expense of society," explains Howard Nusbaum, a neuroscientist at the University of Chicago—leading more and more people to consider how our concepts of reasoning could be extended beyond the traditional definitions of intelligence. Thanks to this wave of attention, we have seen the foundation of new institutions designed to tackle the subject head on, such as Chicago's Center for Practical Wisdom, which opened in 2016 with Nusbaum as its head. The study of wisdom now seems to have reached a kind of tipping point, with a series of exciting recent results.

Igor Grossmann, a Ukrainian-born psychologist at the University of Waterloo, Canada, has been at the cutting edge of this new movement. His aim, he says, is to provide the same level of experimental scrutiny—including randomized controlled trials—that we have come to expect from other areas of science, like medicine. "You're going to need that baseline work before you can go and convince people that 'if you do this it will solve all your problems,'" he told me during an interview at his Toronto apartment. For this reason, he calls the discipline "evidence-based wisdom"—in the same way that we now discuss "evidence-based medicine."

Grossmann's first task was to establish a test of wise reasoning, and to demonstrate that it has real-world consequences that are independent of general intelligence, education, and professional expertise. He began by examining various philosophical definitions of wisdom, which he broke down into six specific principles of thinking. "I guess you would call them metacognitive components—various aspects of knowledge and cognitive processes that can guide you toward a more enriched, complex understanding of a situation," he said.

As you would hope, this included some of the elements of reasoning that we have already examined, including the ability to "consider the perspectives of the people involved in the conflict," which takes into consideration your ability to seek and absorb information that contradicts your initial view; and "recognizing the ways in which the conflict might unfold," which involves the counterfactual thinking that Sternberg had studied in his measures of creative intelligence, as you try to imagine the different possible scenarios.

But his measure also involved some elements of reasoning that we haven't yet explored, including an ability to "recognize the likelihood of change," "search for a compromise" and "predict conflict resolution."

Last, but not least, Grossmann considered intellectual humility—an awareness of the limits of our knowledge, and inherent uncertainty in our judgment; essentially, seeing inside your bias blind spot. It's the philosophy that had guided Socrates more than two millennia ago, and which also lay at the heart of Franklin's speech at the signing of the US Constitution.

Having identified these characteristics, Grossmann asked his participants to think out loud about various dilemmas—from newspaper articles concerning international conflicts to a syndicated "Dear Abby" agony aunt column about a family disagreement, while a team of colleagues scored them on the various traits.

To get a flavor of the test, consider the following dilemma:

> Dear Abby,
>
> My husband, "Ralph," has one sister, "Dawn," and one brother, "Curt." Their parents died six years ago, within months of each other. Ever since, Dawn has once a year mentioned buying a headstone for their parents. I'm all for it, but Dawn is determined to spend a bundle on it, and she expects her brothers to help foot the bill. She recently told me she had put $2,000 aside to pay for it. Recently Dawn called to announce that she had gone ahead, selected the design, written the epitaph and ordered the headstone. Now she expects Curt and Ralph to pay "their share" back to her.

> She said she went ahead and ordered it on her own because she has been feeling guilty all these years that her parents didn't have one. I feel that since Dawn did this all by herself, her brothers shouldn't have to pay her anything. I know that if Curt and Ralph don't pay her back, they'll never hear the end of it, and neither will I.

The response of a participant scoring low on humility looked something like this:

> I think the guys probably end up putting their share in . . . or she will never hear the end of it. I *am sure* they have hard feelings about it, but *I am sure* at the end they will break down and help pay for it.[12]

The following response, which acknowledges some crucial but missing information, earned a higher score for humility:

> Dawn apparently is impatient to get this done, and the others have been dragging it out for 6 years or at least nothing's been done for 6 years. *It doesn't say how much she finally decided would be the price . . . I don't know that that's how it happened*, just that that seems the reasonable way for them to go about it. *It really depends on the personalities of the people involved, which I don't know.*

Similarly, for perspective taking, a less sophisticated response would examine just one point of view:

> I can imagine that it was a sour relationship afterward because let's just say that Kurt and Ralph decided not to go ahead and pay for the headstone. Then it is going to create a gap of communication between her sister and her brothers.

A wiser response instead begins to look more deeply into the potential range of motives:

> Somebody might believe that we need to honor parents like this. Another person might think there isn't anything that needs to be

> done. Or another person might not have the financial means to do anything. Or it could also mean that it might not be important to the brothers. It often happens that people have different perspectives on situations important to them.

The high scorer could also see more possibilities for the way the conflict might be resolved:

> I would think there would probably be some compromise reached, that Kurt and Ralph realize that it's important to have some kind of headstone, and although Dawn went ahead and ordered it without them confirming that they'd pitch in, they would probably pitch in somehow, even if not what she wanted ideally. But hopefully, there was some kind of contribution.

As you can see, the responses are very conversational—they don't demand advanced knowledge of philosophical principles, say—but the wiser participants are simply more willing to think their way around the nuances of the problem.

After the researchers had rated the participants' thinking, Grossmann compared these scores to different measures of well-being. The first results, published in 2013 in the *Journal of Experimental Psychology*, found that people with higher scores for wise reasoning fared better in almost every aspect of life: they were more content and less likely to suffer depression, and they were generally happier with their close relationships.

Strikingly, they were also slightly less likely to die during a five-year follow-up period, perhaps because their wiser reasoning meant they were better able to judge the health risks of different activities, or perhaps because they were better able to cope with stress. (Grossmann emphasises that further work is needed to replicate this particular finding, however.)

Crucially, the participants' intelligence was largely unrelated to their wise reasoning scores, and had little bearing on any of these measures of health and happiness.[13] The idea that "I am wise because I know that I know nothing" may have become something of a cliché, but it is still rather remarkable that qualities such as your intellectual humility and capacity to understand other people's points of view may predict your well-being *better than your actual intelligence.*

This discovery complements other recent research exploring intelligence, rational decision making, and life outcomes. You may recall, for instance, that Wändi Bruine de Bruin found very similar results, showing that her measure of "decision making competence" was vastly more successful than IQ at predicting stresses like bankruptcy and divorce.[14] "We find again and again that intelligence is a little bit related to wise reasoning—it explains perhaps 5% of the variance, probably less, and definitely not more than that," said Grossmann.

Strikingly, Grossmann's findings also converge with Keith Stanovich's research on rationality. One of Stanovich's sub-tests, for instance, measured a trait called "actively open-minded thinking," which overlaps with the concept of intellectual humility, and which also includes the ability to think about alternative perspectives. How strongly would you agree with the statement that "Beliefs should always be revised in response to new information or evidence," for instance? Or "I like to gather many different types of evidence before I decide what to do"? He found that participants' responses to these questions often proved to be a far better predictor of their overall rationality than their general intelligence—which is reassuring, considering that unbiased decision making should be a key component of wisdom.[15]

Grossmann agrees that a modest level of intelligence will be necessary for some of the complex thinking involved in these tasks. "Someone with severe learning difficulties won't be able to apply these wisdom principles." But beyond a certain threshold, the other characteristics—such as intellectual humility and open-minded thinking—become more crucial for the decisions that truly matter in life.

Since Grossmann published those results, his theories have received widespread acclaim from other psychologists, including a Rising Star Award from the American Psychological Association.[16] His later research has built on those earlier findings with similarly exciting results. With Henri Carlos Santos, for instance, he examined longitudinal data from previous health and well-being surveys that had, by good fortune, included questions on some of the qualities that are important to his definition of wisdom, including intellectual humility and open-mindedness. Sure enough, he found that people who scored more highly on these characteristics at the start of the survey tended to report greater happiness later on.[17]

He has also developed methods that allow him to test a greater number of people. One study asked participants to complete an online diary for nine days, with details about the problems they faced and questionnaires examining their thinking in each case. Although some people consistently scored higher than others, their behavior was still highly dependent on the situation at hand. In other words, even the wisest person may act foolishly in the wrong circumstances.[18]

This kind of day-to-day variation can be seen in personality traits such as extraversion, Grossmann says, as each person's behavior varies from a fixed set point; a mild introvert may still prefer to be quietly alone at work, but then become more gregarious around the people she trusts. Similarly, it's possible that someone may be fairly wise when dealing with a confrontational colleague—but then lose their head when dealing with their ex.

The question is, how can we learn to change that set point?

~

Benjamin Franklin's writings offer anecdotal evidence that wisdom can be cultivated. According to his autobiography, he had been a "disputatious" youth, but that changed when he read an account of Socrates's trial.[19] Impressed by the Greek philosopher's humble method of enquiry, he determined to always question his own judgment and respect other people's, and in his conversation, he refused to use words such as "certainly, undoubtedly, or any others that give the air of positiveness to an opinion." Soon it became a permanent state of mind. "For these fifty years past no one has ever heard a dogmatical expression escape me," he wrote.

The result was the kind of humble and open mind that proves to be so critical for Grossmann's research on evidence-based wisdom. "I find a frank acknowledgment of one's ignorance is not only the easiest way to get rid of a difficulty, but the likeliest way to obtain information, and therefore I practice it," Franklin wrote in 1755, while discussing his confusion over a recent scientific result. "Those who affect to be thought to know everything, and so undertake to explain everything, often remain long ignorant of many things that others could and would instruct them in, if they appeared less conceited."[20]

Unfortunately, the scientific research suggests that good intentions may not be sufficient. A classic psychological study by Charles Lord in the late 1970s found that simply telling people to be "as objective and unbiased as possible" made little to no difference in counteracting the myside bias. When considering arguments for the death penalty, for instance, subjects still tended to come to conclusions that suited their preconceptions and still dismissed the evidence opposing their view, despite Lord's warnings.[21] Clearly, wanting to be fair and objective alone isn't enough; you also need practical methods to correct your blinkered reasoning.

Luckily, Franklin had also developed some of those strategies—methods that psychologists would only come to recognize centuries later.

His approach is perhaps best illustrated through a letter to Joseph Priestley in 1772. The British clergyman and scientist had been offered the job of overseeing the education of the aristocrat Lord Shelburne's children. This lucrative opportunity would offer much-needed financial security, but it would also mean sacrificing his ministry, a position he considered "the noblest of all professions"—and so he wrote to Franklin for advice.

"In the affair of so much importance to you, where-in you ask my advice, I cannot, for want of sufficient premises, counsel you *what* to determine, but if you please, I will tell you *how*," Franklin replied. He called his method a kind of "moral algebra," and it involved dividing a piece of paper in two and writing the advantages and disadvantages on either side—much like a modern pros and cons list. He would then think carefully about each one and assign them a number based on importance; if a pro equaled a con, he would cross them both off the list. "Thus proceeding I find at length where the *balance* lies; and if, after a day or two of farther consideration, nothing new that is of importance occurs on either side, I come to a determination accordingly."[22]

Franklin conceded that the values he placed on each reason were far from scientific, but argued that when "each is thus considered separately and comparatively, and the whole lies before me, I think I can judge better, and am less liable to make a rash step."

As you can see, Franklin's strategy is more deliberative and involved than the quick lists of advantages and disadvantages most of us may scribble in a notebook. Of particular importance is the careful way that he attempts to weigh up each item, and his diligence in suspending his judgment to allow his thoughts to settle. Franklin seems to have been especially aware of our tendency to lean heavily on the reasons that are most easily recalled. As he described in another letter, some people base their decisions on facts that just "happened to be present in the mind," while the best reasons were "absent."[23] This tendency is indeed an important source of bias when we try to reason, which is why it's so important to give yourself the time to wait until all the arguments are laid out in front of you.[24]

Whether or not you follow Franklin's moral algebra to the letter, psychologists have found that deliberately taking time to "consider the opposite" viewpoint can reduce a range of reasoning errors,[25] such as anchoring,[26] and overconfidence,[27] and, of course, the myside bias. The benefits appear to be robust across many different decisions—from helping people to critique dubious health claims[28] to forming an opinion on capital punishment and reducing sexist prejudice.[29] In each case, the aim was to actively argue against yourself, and consider why your initial judgment may be wrong.*[30]

---

* The thirteenth-century philosopher Thomas Aquinas, incidentally, used similar techniques in his own theological and philosophical inquiries. As the philosopher Jason Baehr (a modern champion of intellectual humility, who we'll meet in Chapter 8) points out, Aquinas deliberately argued against his initial hypothesis on any opinion, doing "his best to make these objections as forceful or strong as possible." He then argues against those points with equal force, until eventually his view reaches some kind of equilibrium.

Depending on the magnitude of the decision, you may benefit from undergoing a few iterations of this process, each time reaching for an additional piece of information that you overlooked on your first pass.[31] You should also pay particular attention to the way you consider the evidence opposing your gut instinct, since you may still be tempted to dismiss it out of hand, even after you have acknowledged its existence. Instead, you might ask yourself: "Would I have made the same evaluation, had exactly the same evidence produced results on the other side of the issue?"

Suppose that, like Priestley, you are considering whether to take a new job and you have sought the advice of a friend, who encourages you to accept the offer. You might then ask: "Would I have given the same weight to my friend's judgment had she opposed the decision?"[32] It sounds convoluted, but Lord's studies suggested this kind of approach really can overcome our tendency to dismiss the evidence that doesn't fit our preferred point of view.

You might also try to imagine that someone else will examine your justifications, or even try to present them to a friend or colleague. Many studies have shown that we consider more points of view when we believe that we will need to explain our thinking to others.[33]

~

We can't know if Franklin applied his moral algebra in all situations, but the general principle of deliberate open-minded thinking seems to have dictated many of his biggest decisions. "All the achievements in the public's interest—getting a fire department organized, the streets paved, a library established, schools for the poor supported, and much more—attest to his skill in reading others and persuading them to do what he wanted them to do," writes the historian Robert Middlekauf.[34] "He calculated and measured; he weighed and he assessed. There was a kind of quantification embedded in the process of his thought. . . . This indeed describes what was most rational about Franklin's mind."

This kind of thinking is not always respected, however. Particularly in a crisis, we sometimes revere "strong," single-minded leaders who will stay true to their convictions, and even Franklin was once considered too "soft" to negotiate with the British during the War of Independence. He was later appointed as one of the commissioners, however, and proved to be a shrewd opponent.

And there is some evidence that a more open-minded approach may lie behind many other successful leaders. One analysis, for instance, has examined the texts of UN General Assembly speeches concerning the Middle East conflict from 1947 to 1976, scoring the content for the speakers' consideration and integration of alternative points of view—the kind of open-minded thinking that was so important for Grossmann's measure of wisdom. The researchers found that this score consistently dropped in the periods preceding a war, whereas higher scores seemed to sustain longer intervals of peace.

It would be foolish to read too much into post-hoc analyses—after all, people would naturally become more closed-minded during times of heightened tension.[35] But lab experiments have found that people scoring lower on these measures are more likely to resort to aggressive tactics. And the idea does find further support in an examination of the US's most important political crises in the last 100 years, including John F. Kennedy's handling of the Cuban missile crisis, and Robert Nixon's dealings with the Cambodian invasion of 1970 and the Yom Kippur War of 1973.

Textual analyses of the speeches, letters and official statements made by presidents and their secretaries of state show that the level of open-minded thinking consistently predicted the later outcome of the negotiations, with JFK scoring highly for his successful handling of the Cuban missile crisis, and Dwight Eisenhower for the way he dealt with the two Taiwan Strait conflicts between Mainland China and Taiwan in the 1950s.[36]

In more recent politics, the German Chancellor Angela Merkel is famous for her "analytical detachment," as she famously listens to all perspectives before making a decision; one senior government official describes her as "the best analyst of any given situation that I could imagine."

The Germans have even coined a new word—*merkeln* (to Merkel)—that captures this patient, deliberative stance, though it's not always meant flatteringly, since it can also reflect frustrating indecision.[37] "I am regarded as a permanent delayer sometimes," she has said herself, "but I think it is essential and extremely important to take people along and really listen to them in political talks." And it has served her well, helping her to remain one of the longest-serving European leaders despite some serious economic crises.

If we recall the idea that many intelligent people are like a car speeding along the road without guidance or caution, then Merkel, Eisenhower, and Franklin represent patient, careful drivers: despite their formidable engines, they know when to hit the brakes and check the terrain before deciding on their route.[38]

~

Franklin's moral algebra is just one of many potential ways to cultivate wisdom, and further insights come from a phenomenon known as Solomon's Paradox, which Grossmann named after the legendary king of Israel in the tenth century BC.

According to biblical accounts, God appeared to Solomon in a dream and offered to give him a special gift at the start of his reign. Rather than choosing wealth, honor, or longevity, he chose wisdom of judgment. His insight was soon put to the test when two harlots appeared before him, both claiming to be the mother of a boy. Solomon ordered for the child to be cut in two—knowing that the true mother would rather renounce her claim than see her son killed. The decision is often considered the epitome of impartial judgment—and people soon traveled from across the land to receive his counsel. He led the land to riches and built Jerusalem's Temple.

Yet Solomon is said to have struggled to apply his famously wise judgment in his personal life, which was ruled by intemperate passions. Despite being the chief Jewish priest, for instance, he defied the Torah's commandments by taking a thousand wives and concubines, and he amassed huge personal wealth. He became a ruthless and greedy tyrant, and was so embroiled in his affairs that he neglected to educate his son and prepare him for power. The kingdom ultimately descended into chaos and war.[39]

Three millennia later, Grossmann has found this same "asymmetry" in his own tests of wisdom. Like Solomon, many people reason wisely about other people's dilemmas, but struggle to reason clearly about their own issues, as they become more arrogant in their opinions, and less able to compromise—another form of the bias blind spot.[40] These kinds of errors seem to be a particular problem when we feel threatened, triggering so-called "hot" emotional processing that is narrow and closed-minded.

The good news is that we can use Solomon's Paradox to our advantage by practicing a process called "self-distancing." To get a flavor of its power, think of a recent event that made you feel angry. Now "take a few steps back," almost as if you were watching yourself from another part of the room or on a cinema screen, and describe the unfolding situation to yourself. How did you feel?

In a series of experiments, Ethan Kross at the University of Michigan has shown that this simple process encourages people to take a more reflective attitude toward their problems—using "cool" rather than "hot" processing. He found, for instance, that they were more likely to describe the situation with more neutral words, and they began to look for the underlying reasons for their discontent, rather than focusing on the petty details.[41]

Consider these two examples. The first is from an "immersed," first-person perspective.

> "I was appalled that my boyfriend told me he couldn't connect with me because he thought I was going to hell. I cried and sat on the floor of my dorm hallway and tried to prove to him that my religion was the same as his . . ."

And the second is from the distanced viewpoint:

> "I was able to see the argument more clearly. . . . I initially empathized better with myself but then I began to understand how my friend felt. It may have been irrational but I understand his motivation."

You can see how the event became less personal, and more abstract, for the second participant—and he or she began to look beyond their own experience to understand the conflict.

Kross emphasises that this is not just another form of avoidance, or suppression. "Our conception was not to remove them from the event but to give them a small amount of distance, hold them back a little bit, and then allow them to confront the emotion from a healthier stance," he told me in an interview. "When you do this from an immersed perspective, people tend to focus on what happened to them. Distancing allows them to shift into this meaning-making mode where they put the event into a broader perspective and context."

He has since repeated the finding many times, using different forms of self-distancing. You may imagine yourself as a fly on the wall, for instance, or a well-intentioned observer. Or you may try to imagine your older, wiser self looking back at the event from the distant future. Simply talking about your experiences in the third person ("David was talking to Natasha, when . . .") can also bring about the necessary change of perspective.

Kross points out that many people naturally self-distance to process unpalatable emotions. He points to an interview in which the basketball player LeBron James described his choice to leave the Cleveland Cavaliers (who had nurtured his career) and move to the Miami Heat. "One thing I didn't want to do was make an emotional decision. I wanted to do what's best for *LeBron James* and to do what makes *LeBron James* happy." Malala Yousafzai, meanwhile, used a similar approach to bolster her courage against the Taliban. "I used to think that the Tali[ban] would come and he would just kill me. But then I said [to myself], if he comes, what would you do Malala? Then I would reply to myself, Malala just take a shoe and hit him."

People who spontaneously take a new perspective in this way enjoy a range of benefits, including reduced anxiety and rumination.[42] Adopting that distanced perspective even helped one group of study participants to confront one of the most feared events in modern life: public speaking. Using self-distancing as they psyched themselves up for a speech, they showed fewer physiological signs of threat, and reported less anxiety, than a control group taking the immersed, first-person perspective. The benefits were also visible to observers judging their talks, too, who thought they gave more confident and powerful speeches.[43]

In each case, self-distancing had helped the participants to avoid that self-centered "hot" cognition that fuels our bias, so that their thinking was no longer serving their anger, fear, or threatened ego. Sure enough, Grossmann has found that self-distancing resolved Solomon's Paradox when thinking about personal crises (such as an unfaithful partner), meaning that people were more humble and open to compromise, and more willing to consider the conflicting viewpoints.[44] "If you become an observer, then right away you get into this inquisitive mode and you try to make sense of the situation," Grossmann told me. "It almost always co-occurs with being intellectually humble, considering different perspectives and integrating them together."

And that may have a serious impact on your relationships. A team led by Eli Finkel at Northwestern University tracked 120 married couples over a period of two years. The initial arc of their relationships was not promising: over the first twelve months, most of the couples faced a downward spiral in their relationship satisfaction, as disappointment and resentments started to build. After a year, however, Finkel gave half of the couples a short course on self-distancing—such as imagining a dispute through the eyes of a more dispassionate observer.

Compared to typical relationship counseling, it was a tiny step—the lesson in self-distancing lasted about twenty minutes in total. But it transformed the couples' love stories, resulting in greater intimacy and trust over the following year, as they constructively worked through their differences. The control group, in contrast, continued their steady decline for the next year, as resentment continued to build.[45]

These are highly intimate problems, but taking a distant viewpoint also seems to remedy bias on less personal subjects. When told to imagine how citizens in other countries would view forthcoming elections, for instance, Grossmann's participants became more open-minded to conflicting views. After the experiment, he found that they were also more likely to take up an offer to sign up for a bipartisan discussion group—offering further, objective evidence that they were now more open to dialogue as a result of the intervention.[46]

As the research evolves, Grossmann has now started to examine the conditions of the effect more carefully, so that he can find even more effective self-distancing techniques to improve people's reasoning. One particularly potent method involves imagining that you are explaining the issue to a twelve-year-old child. Grossmann speculates that this may prime you to be more protective, so that you avoid any bias that could sway their young and naïve mind.[47]

His team call this phenomenon the "Socrates Effect"—the humble, Greek philosopher correcting the egocentric passions of the mighty Israelite king.

~

If you still doubt that these principles will help you make better decisions, consider the achievements of Michael Story, a "super-forecaster" whose talents first came to light through the Good Judgment Project—a US government-funded initiative to improve its intelligence program.

The Good Judgment Project was the brainchild of Philip Tetlock, a political scientist who had already caused shockwaves among intelligence analysts. Whenever we turn on the TV news or read a newspaper, we meet commentators who claim to know who will win an election or if a terrorist attack is imminent; behind closed doors, intelligence analysts may advise governments to go to war, direct NGOs' rescue efforts or advise banks on the next big merger. But Tetlock had previously shown that these professionals often perform no better than if they had been making random guesses—and many performed consistently worse.

Later research has confirmed that their rapid, intuitive decision making makes many intelligence analysts more susceptible to biases such as framing—scoring worse than students on tests of rationality.[48]

It was only after the US-led invasion of Iraq in 2003—and the disastrous hunt for Saddam Hussein's "Weapons of Mass Destruction"—

that the US intelligence services finally decided to take action. The result was the founding of a new department—Intelligence Advanced Research Projects Activity. They eventually agreed to fund a four-year tournament, beginning in 2011, allowing researchers to arrange the participants in various groups and test their strategies.

Example questions included: "Will North Korea detonate a nuclear device before the end of the year?" "Who will come top of the 2012 Olympics medals table?" And, "How many additional countries will report cases of the Ebola virus in the next eight months?" In addition to giving precise predictions on these kinds of events, the forecasters also had to declare their confidence in their judgments—and they would be judged extra harshly if they were overly optimistic (or pessimistic) about their predictions.

Tetlock's team was called the Good Judgment Project, and after the first year he siphoned off the top 2 percent, whom he called the "super-forecasters," to see if they might perform better in teams than by themselves.

Michael joined the tournament midway through the second year, and he quickly rose to be one of the most successful. Having worked in various jobs, including documentary filmmaking, he had returned to academia for a Master's degree, when he saw an ad for the tournament on an economics blog. The idea of being able to test and quantify his predictions instantly appealed.

Michael can still remember meeting other "supers" for the first time. "There are loads of weird little things about us that are very similar," he told me; they share an inquisitive, hungry mind with a thirst for detail and precision, and this was reflected in their life decisions. One of his friends compared it to the ending of *ET*, "where he goes back to his home planet, and he meets all the other ETs."

Their observations tally with Tetlock's more formal investigations. Although the super-forecasters were all smart on measures of general intelligence, "they did not score off-the-charts high and most fall well short of so-called genius territory," Tetlock noted. Instead, he found that their success depended on many other psychological traits—including the kind of open-minded thinking, and the acceptance of uncertainty, that was so important in Grossmann's research.

"It's being willing to acknowledge that you have changed your mind many times before—and you'll be willing to change your mind many times again," Michael told me. The super-forecasters were also highly precise with their declarations of confidence—specifying 22 percent certainty, as opposed to 20 percent, say—which perhaps reflects an overall focus on detail and precision.

Tetlock had already seen signs of this in his earlier experiments, finding that the worst pundits tended to express themselves with the most confidence, while the best performers allowed more doubt to creep into their language, "sprinkling their speech with transition markers such as 'however,' 'but,' 'although,' and 'on the other hand.'"

Remember Benjamin Franklin's determination to avoid "certainly, undoubtedly, or any other [phrases] that give the air of positiveness to an opinion"? More than two hundred years later, the super-forecasters were again proving exactly the same point: it pays to admit the limits of your knowledge.

In line with Grossmann's research, the super-forecasters also tended to look for outside perspectives; rather than getting stuck in the fine details of the specific situation at hand, they would read widely and look for parallels with other (seemingly unconnected) events. Someone investigating the Arab Spring, for instance, may look beyond Middle Eastern politics to see how similar revolutions had played out in South America.

Interestingly, many of the super-forecasters—including Michael—had lived and worked abroad at some point in their life. Although this may have just been a coincidence, there is some good evidence that a deep engagement with other cultures can promote open-minded thinking, perhaps because it demands that you temporarily put aside your preconceptions and adopt new ways of thinking.[49]

The most exciting result, however, was the fact that these skills improved with training. With regular feedback, many people saw their accuracy slowly climbing over the course of the tournament. The participants also responded to specific lessons. An hour-long online course to recognize cognitive bias, for instance, improved the forecasters' estimates by around 10 percent over the following year.

Often, the simplest way to avoid bias was to start out with a "base rate": examining the average length of time it takes for any dictator to fall from power, for instance—before you then begin to readjust the estimate. Another simple strategy was to examine the worst- and best-case scenarios for each situation, offering some boundaries for your estimates.

Overall, the super-forecasters provided the perfect independent demonstration that wise decision making relies on many alternative thinking styles, besides those that are measured on standard measures of cognitive ability. As Tetlock puts it in his book *Superforecasting*: "A brilliant puzzle-solver may have the raw material for forecasting, but if he doesn't also have an appetite for questioning basic, emotionally charged beliefs, he will often be at a disadvantage relative to a less intelligent person who has a greater capacity for self-critical thinking."[50]

Grossmann says that he has only just come to appreciate these parallels. "I think there is quite a bit of convergence in those ideas," he told me.

Michael now works for a commercial spin-off, Good Judgment Inc., which offers courses in these principles, and he confirms that performance can improve with practice and feedback. However you perform, it's important not to fear failure. "You learn by getting it wrong," Michael told me.

~

Before I finished my conversation with Grossmann, we discussed one final, fascinating experiment that took his wise reasoning tests to Japan.

As in Grossmann's previous studies, the participants answered questions about news articles and agony aunt columns, and were then scored on the various aspects of wise reasoning, such as intellectual humility, the ability to take on board another viewpoint, and their ability to suggest a compromise.

The participants ranged from twenty-five to seventy-five years old, and in the US, wisdom grew steadily with age. That's reassuring: the more we see of life, the more open-minded we become. And it's in line with some of the other measures of reasoning, such as Bruine de Bruin's "adult decision-making competence scale," in which older people also tend to score better.

But Grossmann was surprised to find that the scores from Tokyo took a completely different pattern. There was no steep increase in age, because the younger Japanese were already as wise as the oldest Americans. Somehow, by the age of twenty-five, they had already absorbed the life lessons that only come to the Americans after decades more experience.[51]

Reinforcing Grossmann's finding, Emmanuel Manuelo, Takashi Kusumi, and colleagues recently surveyed students in the Japanese cities of Okinawa and Kyoto, and Auckland in New Zealand, on the kinds of thinking that they thought were most important at university. Although all three groups recognized the value of having an open-minded outlook, it's striking that the Japanese students referred to some specific strategies that sound very much like self-distancing. One student from Kyoto emphasised the value of "thinking from a third person's point of view," for instance, while a participant in Okinawa said it was important to "think flexibly based on the opposite opinion."[52]

What could explain these cultural differences? We can only speculate, but many studies have suggested that a more holistic and interdependent view of the world may be embedded in Japanese culture; Japanese people are more likely to focus on the context and to consider the broader reasons for someone's actions, and less likely to focus on the "self."[53]

Grossmann points to ethnographic evidence showing that children in Japan are taught to consider others' perspectives and acknowledge their own weaknesses from a young age. "You just open an elementary school textbook and you see stories about these characters who are intellectually humble, who think of the meaning of life in interdependent terms."

Other scholars have argued that this outlook may also be encoded in the Japanese language itself. The anthropologist Robert J. Smith noted that the Japanese language demands that you encode people's relative status in every sentence, while the language lacks "anything remotely resembling the personal pronoun." Although there are many possible ways to refer to yourself, "none of the options is clearly dominant," particularly among children. "With overwhelming frequency, they use no self-referent of any kind."

Even the pronunciation of your own name changes depending on the people with whom you are speaking. The result, Smith said, is that self-reference in Japan is "constantly shifting" and "relational" so that "there is no fixed center from which the individual asserts a noncontingent existence."[54] Being forced to express your actions in this way may naturally promote a tendency for self-distancing.

Grossmann has not yet applied his wise reasoning tests to other countries, but converging evidence would suggest that these differences should be considered part of broader geographical trends.

Thanks, in part, to the practical difficulties inherent in conducting global studies, psychologists once focused almost entirely on Western populations, with the vast majority of findings emerging from US university students—highly intelligent, often middle-class people. But during the last ten years, they have begun to make a greater effort to compare the thinking, memory, and perception of people across cultures. And they are finding that "Western, Educated, Industrialized, Rich, Democratic" (WEIRD, for short) regions like North America and Europe score higher on various measures of individualism and the egocentric thinking that appears to lie behind our biases.

In one of the simplest "implicit" tests, researchers ask participants to draw a diagram of their social network, representing their family and friends and their relationships to each other. (You could try it for yourself, before you read on.)

In WEIRD countries like the US, people tend to represent themselves as bigger than their friends (by about 6 mm on average) while people from China or Japan tend to draw themselves as slightly smaller than the people around them.[55] This is also reflected in the words they use to describe themselves: Westerners are more likely to describe their own personality traits and achievements, while East Asian people describe their position in the community. This less individualistic, more "holistic" way of viewing the world around us can also be seen in India, the Middle East, and South America,[56] and there is some emerging evidence that people in more interdependent cultures find it easier to adopt different perspectives and absorb other people's points of view—crucial elements of wisdom that would improve people's thinking.[57]

Consider measures of overconfidence, too. As we have seen, most WEIRD participants consistently overestimate their abilities: 94 percent of American professors rate themselves as "better than average," for instance, and 99 percent of car drivers think they are more competent than the average.[58] Yet countless studies have struggled to find the same tendency in China, Korea, Singapore, Taiwan, Mexico, or Chile.[59] Of course, that's not to say that everyone in these countries will *always* be humble, wise thinkers; it almost certainly depends on the context, as people naturally flip between different ways of thinking. And the general characteristics may be changing over time. According to one of Grossmann's recent surveys, individualism is rising across the globe, even in populations that traditionally showed a more interdependent outlook.[60]

Nevertheless, we should be ready to adopt the more realistic view of our own abilities that is common in East Asian and other cultures, as it could directly translate to a smaller "bias blind spot," and better overall reasoning.

~

We have now seen how certain dispositions—particularly intellectual humility and actively open-minded thinking—can help us to navigate our way around the intelligence trap. And with Franklin's moral algebra and self-distancing, we have two solid techniques that can immediately improve our decision making. They aren't a substitute for greater intelligence or education, but they help us to apply that brainpower in a less biased fashion, so that we can use it more fruitfully while avoiding any intellectual landmines.

The science of evidence-based wisdom is still in its infancy, but over the next few chapters we will explore convergent research showing how cutting-edge theories of emotion and self-reflection can reveal further practical strategies to improve our decision making in high-stakes environments. We'll also examine the ways that an open-minded, humble attitude, combined with sophisticated critical thinking skills, can protect us from forming dangerous false beliefs and from "fake news."

~

Benjamin Franklin continued to embody intellectual humility to the very end. The signing of the Constitution in 1787 was his final great act, and he remained content with his country's progress. "We have had a most plentiful year for the fruits of the earth, and our people seem to be recovering fast from the extravagant and idle habits which the war had introduced, and to engage seriously in the contrary habits of temperance, frugality, and industry, which give the most pleasing prospects of future national felicity," he wrote to an acquaintance in London in 1789.[61]

In March 1790, the theologian Ezra Stiles probed Franklin about his own beliefs in God and his chances of an afterlife. He replied: "I have, with most of the Dissenters in England, some doubts as to [Jesus's] divinity, *though it is a question I do not dogmatize upon*, having never studied it, and think it needless to busy myself with it now, when I expect soon an opportunity of knowing the truth with less trouble.

"I shall only add, respecting myself, that, having experienced the goodness of that Being in conducting me prosperously through a long life, I have no doubt of its continuance in the next, though without the smallest conceit of meriting such goodness."[62] He died little more than a month later.

5

Your emotional compass: *The power of self-reflection*

As he hungrily ate his burger and fries, Ray had already begun to sketch out his business plan. The fifty-two-year-old salesman was not a gambling man, but when he got this intense visceral feeling in his "funny bone," he knew he had to act—and he had never before felt an intuition this strong.

Those instincts hadn't led him astray yet. He had made his way from playing jazz piano in bars and bordellos to a successful career in the paper cup industry, becoming his company's most successful salesman. Then, soon after the Second World War, he had seen the potential in milkshake mixers, and he was now making a tidy sum selling them to diners.

But his mind was always open to new possibilities. "As long as you're green you're growing; as soon as you're ripe you start to rot," he liked to say. And although his body may have been telling him otherwise—he had diabetes and the beginnings of arthritis—he still felt as green as people half his age.

So when he noticed that new clients were flocking to him on the recommendation of one particular hamburger joint, owned by two brothers in San Bernardino, California, he knew he had to take a look. What was so special about this one outlet that had inspired so many others to pay out for a better shake maker?

Entering the premises, he was struck first by the cleanliness of the operation: everyone was dressed in pristine uniforms, and unlike the typical roadside restaurant, it wasn't swarming with flies. And although the menu was limited, the service was quick and efficient.

Each step of the food production was stripped down to its essence, and by paying with the order, you could come in and go out without even having to wait around tipping waitresses. Then there was the taste of the French fries, cut from Idaho potatoes that were cooked to perfection in fresh oil, and the burgers, fried all the way through with a slice of cheese on one side. You could, the sign outside read, "buy 'em by the bag."

Ray had never been to a burger joint like it; it was somewhere he would have happily taken his wife and children. And he saw that the operation could easily be upscaled. His excitement was visceral; he was "wound up like a pitcher with a no-hitter going." He knew he had to buy the rights to franchise the operation and spread it across America.[1]

Within the next few years, Ray would risk all his savings to buy out the two brothers who owned it. He would keep the emblem of its golden arches, though, and despite the acrimonious split, the brothers' name—McDonald—would still be emblazoned on every restaurant.

His lawyers apparently thought he was mad; his wife's reaction was so negative that they got divorced. But Ray was never in any doubt. "I felt in my funny bone that it was a sure thing."[2]

~

History may have proven Ray Kroc's instincts correct; McDonald's serves nearly 70 million customers *every day*. In light of the science of dysrationalia, however, it's natural to feel more than a little skeptical of a man who gambled everything on the whims of his funny bone.

Surely instinctual reasoning like this is the antithesis of Franklin's slow-and-careful moral algebra and Igor Grossmann's study of evidence-based wisdom? We've seen so many examples of people who have followed their hunches to their detriment; Kroc would seem to be the exception who proves the rule. If we want to apply our intelligence more rationally, we should always try to avoid letting our emotions and gut feelings rule our actions in this way.

This would be a grave misunderstanding of the research, however. Although our gut reactions are undoubtedly unreliable, and overconfidence in those feelings will lead to dysrationalia, our emotions and intuitions can also be valuable sources of information, directing our thinking in impossibly complex decisions and alerting us to details that have been accidentally overlooked through conscious deliberation.

The problem is that most people—including those with high general intelligence, education, and professional expertise—lack the adequate self-reflection to interpret the valuable signals correctly and identify the cues that are going to lead them astray. According to the research, bias doesn't come from intuitions and emotions per se, but from an inability to recognize those feelings for what they really are and override them when necessary; we then use our intelligence and knowledge to justify erroneous judgments made on the basis of them.

Cutting-edge experiments have now identified exactly what skills are needed to analyze our intuitions more effectively, suggesting yet more abilities that are not currently recognized in our traditional definitions of intelligence, but which are essential for wise decision making. And it turns out that Kroc's descriptions of his physical "funny-bone feelings" perfectly illustrate this new understanding of the human mind.

The good news is that these reflective skills can be learned, and when we combine them with other principles of evidence-based wisdom, the results are powerful. These strategies can improve the accuracy of your memories, boost your social sensitivity so that you become a more effective negotiator, and light the spark of your creativity.

By allowing us to de-bias our intuitions, these insights resolve many forms of the intelligence trap, including the curse of expertise that we explored in Chapter 3. And some professions are already taking notice. In medicine, for instance, these strategies are being applied by doctors who hope to reduce diagnostic errors—techniques that could potentially save tens of thousands of lives every year.

~

Like much of our knowledge about the brain's inner workings, this new understanding of emotion comes from the extreme experiences of people who have sustained neurological injury to a specific part of the brain.

In this case, the area of interest is the ventromedial area of the prefrontal cortex, located just above the nasal cavity—which may be damaged through surgery, stroke, infection, or a congenital defect.

Superficially, people with damage to this area appear to emerge from these injuries with their cognition relatively unscathed: they still score well on intelligence tests, and their factual knowledge is preserved. And yet their behavior is nevertheless extremely bizarre, veering between incessant indecision and rash impulsiveness.

They may spend hours deliberating over the exact way to file an office document, for instance, only to then invest all of their savings in a poor business venture or to marry a stranger on a whim. It's as if they simply can't calibrate their thinking to the importance of the decision at hand. Worse still, they appear immune to feedback, ignoring criticism when it comes their way, so they are stuck making the same errors again and again.

"Normal and intelligent individuals of comparable education make mistakes and poor decisions, but not with such systematically dire consequences," the neurologist Antonio Damasio wrote of one of the first known patients, Elliot, in the early 1990s.*[3]

Damasio was initially at a loss to explain why damage to the frontal lobe would cause this strange behavior. It was only after months of observation with Elliot that Damasio uncovered another previously undiscovered symptom that would eventually hold the key to the puzzle: despite the fact that his whole life was unraveling in front of him, Elliot's mood never once faltered from an eerie calmness. What Damasio had originally taken to be a stiff upper lip seemed like an almost complete lack of emotion. "He was not inhibiting the expression of internal emotional resonance or hushing inner turmoil," Damasio later wrote. "He simply did not have any turmoil to hush."

Those observations would ultimately lead Damasio to propose the "somatic marker hypothesis" of emotion and decision making. According to this theory, any experience is immediately processed nonconsciously, and this triggers a series of changes within our body—such as fluctuations in heart rate, a knot in the stomach, or

---

* As in many medical case studies, Damasio has used a pseudonym for his patient.

build-up of sweat on the skin. The brain then senses these "somatic markers" and interprets them according to the context of the situation and its knowledge of emotional states. Only then do we become conscious of how we are feeling.

This process makes evolutionary sense. By continually monitoring and modifying blood pressure, muscle tension, and energy consumption, the brain can prepare the body for action, should we need to respond physically, and maintains its homeostasis. In this way, the somatic marker hypothesis offers one of the best, biologically grounded, theories of emotion. When you feel the rush of excitement flowing to the tips of your fingers, or the unbearable pressure of grief weighing on your chest, it is due to this neurological feedback loop.

Of even greater importance for our purposes, however, the somatic marker hypothesis can also explain the role of intuition during decision making. According to Damasio, somatic markers are the product of rapid nonconscious processing which creates characteristic bodily changes before our conscious reasoning has caught up. The resulting physical sensations are the intuitive feelings we call gut instinct, giving us a sense of the correct choice before we can explain the reasons why.

The ventromedial prefrontal cortex, Damasio proposed, is one of the central hubs that is responsible for creating bodily signals based on our previous experiences, explaining why patients like Elliot failed to feel emotions, and why they would often make bad decisions; their brain damage had cut off their access to the nonconscious information that might be guiding their choices.

Sure enough, Damasio found that people like Elliot failed to show the accompanying physiological responses—such as sweating—when viewing disturbing images (such as a photo of a horrific homicide). To further test his theory, Damasio's team designed an elegant experiment called the Iowa Gambling Task, in which participants are presented with four decks of cards. Each card can come with a small monetary reward, or a penalty, but two of the decks are subtly stacked against the player, with slightly bigger rewards but much bigger penalties. The participants don't initially know this, though: they just have to take a gamble.

For most healthy participants, the body starts showing characteristic changes in response to a particular choice—such as signs of stress when the participant is set to choose the disadvantageous deck—before the player is consciously aware that some decks are stacked for or against them. And the more sensitive someone is to their bodily feelings—a sense called interoception—the quicker they learn how to make the winning choices.

As Damasio expected, brain injury survivors such as Elliot were especially bad at the Iowa Gambling Task, making the wrong choices again and again long after others have started homing in on the right card decks. This was caused by their lack of the characteristic somatic changes before they made their choices. Unlike other players, they did not experience a reliable visceral response to the different decks that would normally warn people from risking huge losses.[4]

You don't need to have endured a brain injury to have lost touch with your feelings, though. Even among the healthy population, there is enormous variation in the sensitivity of people's interoception, a fact that can explain why some people are better at making intuitive decisions than others.

You can easily measure this yourself. Simply sit with your hands by your sides and ask a friend to take your pulse. At the same time, try to feel your own heart in your chest (without actually touching it) and count the number of times it beats; then, after one minute, compare the two numbers.

How did you do? Most people's estimates are off by at least 30 percent,[5] but some reach nearly 100 percent accuracy—and your place on this scale will indicate how you make intuitive decisions in exercises like the Iowa Gambling Task, with the higher scorers naturally gravitating to the most advantageous options.[6]

Your score on the heartbeat counting test can translate to real-world financial success, with one study showing that it can predict the profits made by traders in an English hedge fund, and how long they survived within the financial markets.[7] Contrary to what we might have assumed, it is the people who are most sensitive to their visceral "gut" feelings—those with the most accurate interoception—who made the best possible deals.

Its importance doesn't end there. Your interoceptive accuracy will also determine your social skills: our physiology often mirrors the signals we see in others—a very basic form of empathy—and the more sensitive you are to those somatic markers, the more sensitive you will be of others' feelings too.[8]

Tuning into those signals can also help you to read your memories. It is now well known that human recall is highly fallible, but somatic markers signal the confidence of what you think you know[9]—whether you are certain or simply guessing. And a study from Keio University in Tokyo found they can also act as reminders when you need to remember to do something in the future—a phenomenon known as prospective memory.[10]

Imagine, for instance, that you are planning to call your mother in the evening to wish her a happy birthday. If you have more attuned interoception, you might feel a knot of unease in your stomach during the day, or a tingling in your limbs, that tells you there's *something* you need to remember, causing you to rack your brain until you recall what it is. Someone who was less aware of those body signals would not notice those physiological reminders and would simply forget all about them.

Or consider a TV quiz show like *Who Wants to Be a Millionaire*. Your success will undoubtedly depend on your intelligence and general knowledge, but your sensitivity to somatic markers will also determine whether you are willing to gamble it all on an answer you don't really know, or whether you can correctly gauge your uncertainty and decide to use a lifeline.

In each case, our nonconscious mind is communicating, through the body, something that the conscious mind is still struggling to articulate. We talk about "following the heart" when we are making important life choices—particularly in love—but Damasio's somatic marker hypothesis shows that there is a literal scientific truth to this romantic metaphor. Our bodily signals are an inescapable element of almost every decision we make, and as the experiences of people like Elliot show, we ignore them at our peril.

When Kroc described the uncanny feeling in his funny bone and the sense of being "wound up like a pitcher," he was almost certainly tapping into somatic markers generated by his nonconscious mind, based on a lifetime's sales experience.

Those feelings determined who he recruited, and who he fired. It was the cause of his decision to first buy into the McDonald's franchise and after their relationship had turned sour, it led him to buy out the brothers. Even his choice to keep the burger bar's original name—when he could have saved millions by starting his own brand—was put down to his gut instincts. "I had a strong intuitive sense that the name McDonald's was exactly right."[11]

Kroc's descriptions offer some of the most vivid examples of this process, but he is far from alone in this. In creative industries, in particular, it's difficult to imagine how you could judge a new idea purely analytically, without some instinctual response.

Consider Coco Chanel's descriptions of her nose for new designs. "Fashion is in the air, born upon the wind. One intuits it. It is in the sky and on the road." Or Bob Lutz, who oversaw the construction of Chrysler's iconic Dodge Viper that helped save the company from ruin in the 1990s. Despite having no market research to back up his choice, he knew that the sports car—way beyond the price range of Chrysler's usual offerings—would transform the company's somewhat dour image. "It was this subconscious visceral feeling . . . it just felt right," he says of his decision to pursue the radical new design.[12]

Damasio's theory, and the broader work on interoception, gives us a strong scientific foundation to understand where those visceral feelings come from and the reasons that some people appear to have more finely tuned intuitions than others.

This cannot be the whole story, however. Everyday experience would tell us that for every Kroc, Chanel, or Lutz, you will find someone whose intuitions have backfired badly, and to make better decisions we still need to learn how to recognize and override those deceptive signals. To do so, we need two additional elements to our emotional compass.

Lisa Feldman Barrett, a psychologist and neuroscientist at Northeastern University in Boston, has led much of this work, exploring both the ways our moods and emotions can lead us astray and potential ways to escape those errors. As one example, she recalls a day at graduate school when a colleague asked her out on a date. She didn't really feel attracted to him, but she'd been working hard and felt like a break, so she agreed to go to the local coffee shop, and as they chatted, she felt flushed and her stomach fluttered—the kinds of somatic markers that you might expect to come with physical attraction. Perhaps it really was love?

By the time she'd left the coffee shop, she had already arranged to go on another date, and it was only when she walked into her apartment and vomited that she realized the true origins of those bodily sensations: she had caught the flu.[13]

The unfortunate fact is that our somatic markers are messy things, and we may accidentally incorporate irrelevant feelings into our interpretations of the events at hand—particularly if they represent "background feelings" that are only on the fringes of our awareness, but which may nevertheless determine our actions.

If you have a job interview, for instance, you'd better hope it's not raining—studies show that recruiters are less likely to accept a candidate if the weather is bad when they first meet them.[14] When researchers spray the smell of a fart, meanwhile, they can trigger feelings of disgust that sway people's judgments of moral issues.[15] And the joy that comes from a World Cup win can even influence a country's stock market—despite the fact that it has nothing to do with the economy's viability.[16]

In each case, the brain was interpreting those background feelings and responding as if they were relevant to the decision at hand. "Feeling," says Feldman Barrett, "is believing"—a phenomenon called "affective realism."[17]

This would seem to pour cold water on any attempts to use our intuition. But Feldman Barrett has also found that some people are consistently better able to disentangle those influences than others—and it all depends on the words they use to describe their feelings.

Perhaps the best illustration comes from a month-long investigation of investors taking part in an online stock market. Contrary to the popular belief that a "cooler head always prevails"—and in agreement with the study of the traders at the London hedge fund—Feldman Barrett found that the best performers reported the most intense feelings during their investments.

Crucially, however, the biggest winners also used more precise vocabularies to describe those sensations. While some people might use the words "happy" and "excited" interchangeably, for example, these words represented a very specific feeling for some people—a skill that Feldman Barrett calls "emotion differentiation."[18]

It wasn't that the poorer performers lacked the words; they simply weren't as careful to apply them precisely to describe the exact sensations they were feeling; "content" and "joyful" both just meant something pleasant; "angry" or "nervous" described their negative feelings. They seemed not to be noting any clear distinctions in their feelings—and that ultimately impaired their investment decisions.

This makes sense given some of the previous research on affective realism, which had found that the influence of irrelevant feelings due to the weather or a bad smell, say, only lasts as long as they linger below conscious awareness, and their power over our decisions evaporates as soon as the extraneous factors are brought to conscious attention. As a consequence, the people who find it easier to describe their emotions may be more aware of background feelings, and they are therefore more likely to discount them. By pinning a concept on a feeling, it is easier to analyze it more critically and to disregard if it is irrelevant.[19]

The benefits of emotion differentiation don't end there. Besides being more equipped to disentangle the sources of their feelings, people with more precise emotional vocabularies also tend to have more sophisticated ways of regulating their feelings when they threaten to get out of hand. A stock market trader, for instance, would be better able to get back on their feet after a string of losses, rather than sinking into despair or attempting to win it all back with increasingly risky gambles.

Sensible regulation strategies include self-distancing, which we explored in the last chapter, and reappraisal, which involves reinterpreting the feelings in a new light. It could also involve humor—cracking a joke to break the tension—or a change of scene. Perhaps you simply realize that you need to get away from the table and take a deep breath. But whatever strategy you use, you can only regulate those feelings once you have already identified them.

For these reasons, people with poor interoception,[20] and low emotional differentiation, are less likely to keep their feelings under wraps before they get out of hand.* Regulation is therefore the final cog in our emotional compass, and, together, those three interconnected components—interoception, differentiation, and regulation—can powerfully direct the quality of our intuition and decision making.[21]

~

I hope you are now convinced that engaging with your feelings is not a distraction from good reasoning, but an essential part of it. By bringing our emotions to the mind's surface, and dissecting their origins and influence, we can treat them as an additional and potentially vital source of information. They are only dangerous when they go unchallenged.

Some researchers call these skills emotional intelligence, but although the description makes literal sense, I'll avoid that term to reduce confusion with some of the more questionable EQ tests that we discussed in Part 1. Instead, I'll describe them as reflective thinking, since they all, in some ways, involve turning your awareness inward to recognize and dissect your thoughts and feelings.

Like the strategies that we explored in the last chapter, these abilities shouldn't be seen as some kind of rival to traditional measures of intelligence and expertise, but as complementary behaviors that ensure we apply our reasoning in the most productive way

* It is risky to read too much into Kroc's autobiography, *Grinding It Out*. But he certainly seems to describe some sophisticated strategies to regulate his emotions when they get out of hand, which he claimed to have picked up earlier in his career. As he put it (pp. 61–2): "I worked out a system that allowed me to turn off nervous tension and shut out nagging questions. . . . I would think of my mind as being a blackboard full of messages, most of them urgent, and I practiced imagining a hand with an eraser wiping that blackboard clean. I made my mind completely blank. If a thought began to appear, zap! I'd wipe it out before it could form."

possible, without being derailed by irrelevant feelings that would lead us off track.

Crucially—and this fact is often neglected, even in the psychological literature—these reflective skills also offer some of the best ways of dealing with the very specific cognitive biases that Kahneman and Tversky studied. They protect us from dysrationalia.

Consider the following scenario, from a study by Wändi Bruine de Bruin (who designed one of the decision-making tests that we explored in Chapter 2).

> You have driven halfway to a vacation destination. Your goal is to spend time by yourself—but you feel sick, and you now feel that you would have a much better weekend at home. You think that it is too bad that you already drove halfway, because you would much rather spend the time at home.

What would you do? Stick with your plans, or cancel them?

This is a test of the sunk cost fallacy—and lots of people state that they would prefer not to waste the drive they've already taken. They keep on ruminating about the time they would lose, and so they try in vain to make the best of it—even though the scenario makes it pretty clear that they'll have to spend the vacation in discomfort as a result. Bruine de Bruin, however, has found that this is not true of the people who can think more reflectively about their feelings in the ways that Feldman Barrett and others have studied.[22]

A Romanian study has found similar benefits with the framing effect. In games of chance, for instance, people are more likely to choose options when they are presented as a gain (i.e. 40 percent chance of winning) compared with when they are presented as a loss (60 percent chance of losing)—even when they mean exactly the same thing. But people with more sophisticated emotion regulation are resistant to these labeling effects and take a more rational view of the probabilities as a result.[23]

Being able to reappraise our emotions has also been shown to protect us against motivated reasoning in highly charged political discussions, determining a group of Israeli students' capacity to consider the Palestinian viewpoint during a period of heightened tension.[24]

It should come as little surprise, then, that an emotional self-awareness should be seen as a prerequisite for the intellectually humble, open-minded thinking that we studied in the last chapter. And this is reflected in Igor Grossmann's research on evidence-based wisdom, which has shown that the highest performers on his wise reasoning tests are indeed more attuned to their emotions, capable of distinguishing their feelings in finer detail while also regulating and balancing those emotions so that their passions do not come to rule their actions.[25]

This idea is, of course, no news to philosophers. Thinkers from Socrates and Plato to Confucius have argued that you cannot be wise about the world around you if you do not first know yourself. The latest scientific research shows that this is not some lofty philosophical ideal; incorporating some moments of reflection into your day will help de-bias every decision in your life.

~

The good news is that most people's reflective skills naturally improve over the course of their lifetime; in ten years' time you'll probably be slightly better equipped to identify and command your feelings than you are today.

But are there any methods to accelerate that process?

One obvious strategy is mindfulness meditation, which trains people to listen to their body's sensations and then reflect on them in a nonjudgmental way. There is now strong evidence that besides its many, well-documented health benefits, regular practice of mindfulness can improve each element of your emotional compass—interoception, differentiation and regulation—meaning that it is the quickest and easiest way to de-bias your decision making and hone your intuitive instincts.[26] (If you are skeptical, or simply tired of hearing about the benefits of mindfulness, bear with me—you will soon see that there are other ways to achieve some of the same effects.)

Andrew Hafenbrack, then at the Institut Européen d'Administration des Affaires in France, was one of the first to document these cognitive effects in 2014. Using Bruine de Bruin's tests, he found that a single fifteen-minute mindfulness session can reduce the incidence of the sunk cost bias by 34 percent. That's a massive reduction—of a very common bias—for such a short intervention.[27]

By allowing us to dissect our emotions from a more detached perspective, mindfulness has also been shown to correct the myside biases that come from a threatened ego,[28] meaning people are less defensive when they are faced with criticism[29] and more willing to consider others' perspectives, rather than doggedly sticking to their own views.[30]

Meditators are also more likely to make rational choices in an experimental task known as the "ultimatum game" that tests how we respond to unfair treatment by others. You play it in pairs, and one partner is given some cash and offered the option to share as much of the money as they want with the other participant. The catch is that the receiver can choose to reject the offer if they think it's unfair—and if that happens, both parties lose everything.

Many people do reject small offers out of sheer spite, even though it means they are ultimately worse off—making it an irrational decision. But across multiple rounds of the game, the meditators were less likely to make this choice. For example, when the opponent offered a measly $1 out of a possible $20, only 28 percent of the nonmeditators accepted the money, compared to 54 percent of the meditators who could set their anger aside to make the rational choice. Crucially, this tolerance correlated with the meditator's interoceptive awareness, suggesting that their more refined emotional processing had contributed to their wiser decision making.[31]

Commanding your feelings in this way would be particularly important during business negotiations, when you need to remain alert to subtle emotional signals from others without getting swept away by strong feelings when the discussions don't go to plan. (Along these lines, a Turkish study has found that differences in emotion regulation can account for 43 percent of the variation in simulated business negotiations.)[32]

Having started meditating to deal with the stress at INSEAD, Hafenbrack says that he has now witnessed all these benefits himself.

"I'm able to disconnect the initial stimulus from my response—and that second or two can make a huge difference in whether you overreact to something or if you respond in a productive way," he told me from the Católica-Lisbon School of Business and Economics in Portugal, where he is now a professor of organizational science. "It makes it easier to think what's really the best decision right now."

If mindfulness really isn't your thing, there may be other ways to hone intuitive instincts and improve your emotion regulation. A series of recent studies has shown that musicians (including string players and singers) and professional dancers have more fine-tuned interoception.[33] The scientists behind these studies suspect that training in these disciplines—which all rely on precise movements guided by sensory feedback—naturally encourages greater bodily awareness.

You don't need to actively meditate to train your emotion differentiation either. Participants in one study were shown a series of troubling images and told to describe their feelings to themselves with the most precise words possible.[34] When shown a picture of a child suffering, for example, they were encouraged to question whether they were feeling sadness, pity, or anger, and to consider the specific differences between those feelings.

After just six trials, the participants were already more conscious of the distinctions between different emotions, and this meant that they were subsequently less susceptible to priming during a moral reasoning task. (By improving their emotion regulation the same approach has, incidentally, also helped a group of people to overcome their arachnophobia.)[35]

The effects are particularly striking, since, like the mindfulness studies, these interventions are incredibly short and simple, with the benefits of a single session enduring more than a week later; even a little bit of time to think about your feelings in more detail will pay lasting dividends.

At the very basic level, you should make sure that you pick apart the tangled threads of feeling, and routinely differentiate emotions such as apprehension, fear, and anxiety; contempt, boredom, and disgust; or pride, satisfaction, and admiration. But given these findings, Feldman Barrett suggests that we also try to learn new words—or invent our own—to fill a particular niche on our emotional awareness.

Just think of the term "hangry"—a relatively recent entry into the English language that describes the particular irritability when we haven't eaten.[36] Although we don't necessarily need psychological research to tell us that low blood sugar will cause an accompanying dip in your mood and a dangerously short fuse, naming the concept means that we are now more aware of the feeling when it does happen, and better able to account for the ways it might be influencing our thinking.

In his *Dictionary of Obscure Sorrows*, the writer and artist John Koenig shows just the kind of sensitivity that Feldman Barrett describes, inventing words such as "liberosis," the desire to care less about things, and "altschmerz"—a weariness with the same old issues you've always had. According to the scientific research, enriching our vocabulary in this way isn't just a poetic exercise: looking for, and then defining, those kinds of nuances will actually change the way you think in profound ways.[37]

If you are really serious about fine-tuning your emotional compass, many of the researchers also suggest that you spend a few minutes to jot down your thoughts and feelings from the day and the ways they might have influenced your decisions. Not only does the writing process encourage deeper introspection and the differentiation of your feelings, which should naturally improve your intuitive instincts; it also ensures you learn and remember what worked and what didn't, so you don't make the same mistakes twice.

You may believe you are too busy for this kind of reflection, but the research suggests that spending a few minutes in introspection will more than pay for itself in the long run. A study by Francesca Gino at Harvard University, for instance, asked trainees at an IT center in Bangalore to devote just fifteen minutes a day to writing and reflecting on the lessons they had learned, while drawing out the more intuitive elements of their daily tasks. After eleven days, she found that they had improved their performance by 23 percent, compared to participants who had spent the same time actively practicing their skills.[38] Your daily commute may be the obvious period to engage your mind in this way.

~

Comprenez-vous cette phrase? Parler dans une langue étrangère modifie l'attitude de l'individu, le rendant plus rationnel et plus sage!

We will shortly see how reflective thinking can potentially save lives. But if you are lucky enough to be bilingual—or willing to learn—you can add one final strategy to your new decision-making toolkit—called the foreign language effect.

The effect hinges on the emotional resonances within the words we speak. Linguists and writers have long known that our emotional experience of a second language will be very different from that of our mother tongue; Vladimir Nabokov, for instance, claimed to feel that his English was "a stiffish, artificial thing" compared to his native Russian, despite becoming one of the language's most proficient stylists: it simply didn't have the same deep resonance for him.[39] And this is reflected in our somatic markers, like the sweat response: when we hear messages in another language, the emotional content is less likely to move the body.

Although that may be a frustration for writers such as Nabokov, Boaz Keysar at the University of Chicago's Center for Practical Wisdom has shown that it may also offer us another way to control our emotions.

The first experiment, published in 2012, examined the framing effect, using English speakers studying Japanese and French, and Korean speakers learning English. In their native languages, the participants were all influenced by whether the scenarios were presented as "gains" or "losses." But this effect disappeared when they used their second language. Now, they were less easily swayed by the wording and more rational as a result.[40]

The "foreign language effect" has since been replicated many times in many other countries, including Israel and Spain, and with many other cognitive biases, including the "hot hand illusion"—the belief, in sport or gambling, that success at one random event means we are more likely to have similar luck in the future.[41]

In each case, people were more rational when they were asked to reason in their second language, compared with their first. Our thinking may feel "stiffish," as Nabokov put it, but the slight emotional distance means that we can think more reflectively about the problem at hand.[42]

Besides offering this immediate effect, learning another language can improve your emotion differentiation, as you pick up new "untranslatable" terms that help you see more nuance in your feelings. And by forcing you to see the world through a new cultural lens, it can exercise your actively open-minded thinking, while the challenge of grappling with unknown phrases increases your "tolerance of ambiguity," a related psychological measure which means that you are better equipped to cope with feelings of uncertainty rather than jumping to a conclusion too quickly. Besides reducing bias, that's also thought to be essential for creativity; tolerance of ambiguity is linked to entrepreneurial innovation, for instance.[43]

Given the effort involved, no one would advise that you learn a language *solely* to improve your reasoning—but if you already speak one or have been tempted to resuscitate a language you left behind at school, then the foreign language effect could be one additional strategy to regulate your emotions and improve your decision making.

If nothing else, you might consider the way it influences your professional relationships with international colleagues; the language you use could determine whether they are swayed by the emotions behind the statement or the facts. As Nelson Mandela once said: "If you talk to a man in a language he understands, that goes to his head. If you talk to him in his language, that goes to his heart."

~

One of the most exciting implications of the research on emotional awareness and reflective thinking is that it may finally offer a way to resolve the "the curse of expertise." As we saw in Chapter 3, greater experience can lead experts to rely on fuzzy, gist-based intuitions that often offer rapid and efficient decision making, but can also lead to error. The implication might have seemed to be that we would need to lose some of that efficiency, but the latest studies show that there are ways to use those flashes of insight while reducing the needless mistakes.

The field of medicine has been at the forefront of these explorations—for good reason. Currently, around 10–15 percent of initial diagnoses are incorrect, meaning many doctors will make at least one error for every six patients they see. Often these errors can be corrected before harm is done, but it is thought that in US hospitals alone, around one in ten patient deaths—between 40,000 and 80,000 per annum—can be traced to a diagnostic mistake.[44]

Could a simple change of thinking style help save some of those lives? To find out, I met Silvia Mamede in the hubbub of Rotterdam's Erasmus Medical Center. Mamede moved to the Netherlands from Ceará, Brazil, more than a decade ago, and she immediately offers me a strong cup of coffee—"not the watery stuff you normally get here"—before sitting opposite me with a notebook in hand. "You organize your ideas better if you have a pencil and paper," she explained. (Psychological research does indeed suggest that your memory often functions better if you are allowed to doodle as you talk.)[45]

Her aim is to teach doctors to be similarly reflective concerning the ways they make their decision making. Like the medical checklist, which the doctor and writer Atul Gawande has shown to be so powerful for preventing memory failures during surgery, the concept is superficially simple: to pause, think, and question your assumptions. Early attempts to engage "system 2" thinking had been disappointing, however; doctors told to use pure analysis, *in place* of intuition—by immediately listing all the alternative hypotheses, for instance—often performed worse than those who had taken a less deliberative, more intuitive approach.[46]

In light of the somatic marker hypothesis, this makes sense. If you ask someone to reflect too early, they fail to draw on their experience, and may become overly focused on inconsequential information. You are blocking them from using their emotional compass, and so they become a little like Damasio's brain injury patients, stuck in their "analysis paralysis." You can't just use system 1 or system 2—you need to use both.

For this reason, Mamede suggests that doctors note down their gut reaction as quickly as possible; and only then should they analyze the evidence for their gut reaction and compare it to alternative hypotheses. Sure enough, she has since found that doctors can improve their diagnostic accuracy by up to 40 percent by taking this simple approach—a huge achievement for such a small measure. Simply telling doctors to revisit their initial hypothesis—without any detailed instructions on re-examining the data or generating new ideas—managed to boost accuracy by 10 percent, which again is a significant improvement for little extra effort.

Importantly, and in line with the broader research on emotion, this reflective reasoning also reduces the "affective biases" that can sway a doctor's intuition. "There are all these factors that could disturb 'System 1'—the patient's appearance, whether they are rich or poor, the time pressure, whether they interrupt you," she said. "But the hope is that reflective reasoning can make the physician take a step back."

To explore one such factor, Mamede recently tested how doctors respond to "difficult" patients, such as those who rudely question the professionals' decisions. Rather than observing real encounters, which would be difficult to measure objectively, Mamede offered fictional vignettes to a group of general practitioners (family doctors). The text mostly outlined their symptoms and test results, but it also included a couple of sentences detailing their behavior.

Many of the doctors did not even report noticing the contextual information, while others were perplexed at the reasons they had been given these extra details. "They said, 'But this doesn't matter! We are trained to look past that, to not look at the behavior. This should make no difference,'" Mamede told me. In fact, as the research on emotion would suggest, it had a huge impact. For more complex cases, the general practitioners were 42 percent more likely to make a diagnostic error for the difficult patients.[47]

If the doctors were told to engage in the more reflective procedure, however, they were more likely to look past their frustration and give a correct diagnosis. It seems that the pause in their thinking allowed them to gauge their own emotions and correct for their frustration, just as the theories of emotion differentiation and regulation would predict.

Mamede has also examined the availability bias, causing doctors to overdiagnose an illness if it has recently appeared in the media and is already on their mind. Again, she has shown that the more reflective procedure eliminates the error—even though she offered no specific instructions or explanations warning them of that particular bias.[48] "It's amazing, when you see the graphs of these studies. The doctors who weren't exposed to the reports of disease had an accuracy of 71 percent, and the biased ones only had an accuracy of 50 percent. And then, when they reflected they went back to the 70 percent," she told me. "So it completely corrected for the bias."

These are astonishing results for such small interventions, but they all show us the power of greater self-awareness, when we allow ourselves to think more reflectively about our intuitions.

Some doctors may resist Mamede's suggestions; the very idea that, after all their training, something so simple could correct their mistakes is bruising to the ego, particularly when many take enormous pride in the power of their rapid intuition. At conferences, for instance, she will present a case on the projector and wait for the doctors to give a diagnosis. "It's sometimes twenty seconds—they just read four or five lines and they say 'appendicitis,'" she told me. "There is even this joke saying that if the doctor needs to think, leave the room."

But there is now a growing momentum throughout medicine to incorporate the latest psychological findings into the physician's daily practice. Pat Croskerry at Dalhousie University in Canada is leading a critical thinking program for doctors—and much of his advice echoes the research we have explored in this chapter, including, for instance, the use of mindfulness to identify the emotional sources of our decision, and, when errors have occurred, the employment of a "cognitive and affective autopsy" to identify the reasons that their intuition backfired. He also advocates "cognitive inoculation"—using case studies to identify the potential sources of bias, which should mean that the doctors are more mindful of the factors influencing their thinking.

Croskerry is still collecting the data from his courses to see the long-term effects on diagnostic accuracy. But if these methods can prevent just a small portion of those 40,000–80,000 deaths per year, they will have contributed more than a major new drug.[49]

~

Although medicine is leading the way, a few other professions are also coming around to this way of thinking. The legal system, for instance, is notoriously plagued by bias—and in response to this research, the American Judges Association has now issued a white paper that advocated mindfulness as one of the key strategies to improve judicial decision making, while also advising each judge to take a moment to "read the dials" and interrogate their feelings in detail, just as neuroscientists and psychologists such as Feldman Barrett are suggesting.[50]

Ultimately, these findings could change our understanding of what it means to be an expert.

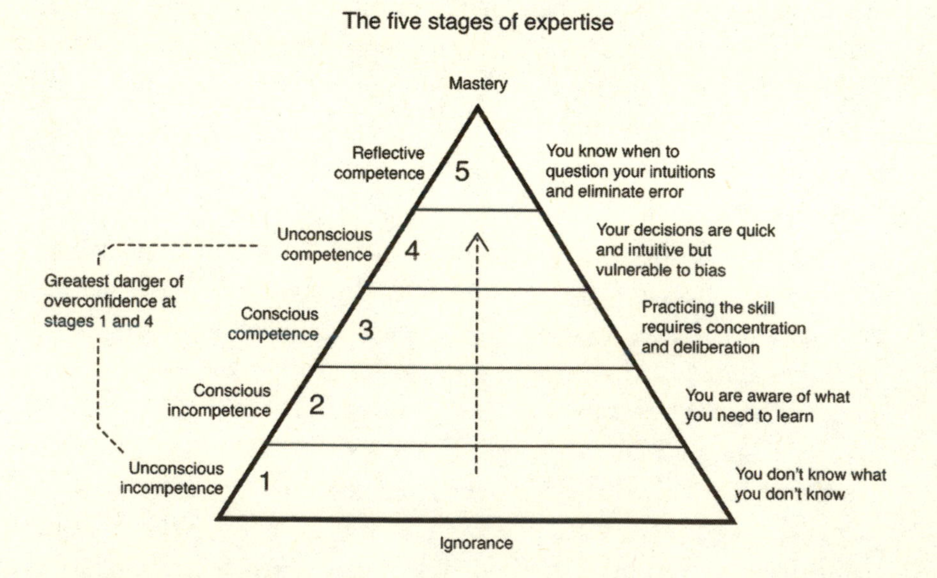

In the past, psychologists had described four distinct stages in the learning curve. The complete beginner is unconsciously incompetent—she does not even know what she doesn't know (potentially leading to the overconfidence of the Dunning-Kruger effect we saw in Chapter 3). After a short while, however, she will understand the skills she lacks, and what she must do to learn them; she is consciously incompetent. With effort, she can eventually become consciously competent—she can solve most problems, but she has to think a lot about the decisions she is making. Finally, after years of training and on-the-job experience, those decisions become second nature—she has reached unconscious competence. This was traditionally the pinnacle of expertise, but as we have seen, she may then hit a kind of "ceiling" where her accuracy plateaus as a result of the expert biases we explored in Chapter 3.[51] To break through that ceiling, we may need one final stage—"reflective competence"—which describes the capacity to explore our feelings and intuitions, and to identify biases before they cause harm.[52]

As Ray Kroc had found in that California diner, intuition can be a powerful thing—but only once we know how to read those funny-bone feelings.

6

A bullshit detection kit: *How to recognize lies and misinformation*

If you were online at the turn of the new millennium, you may remember reading about the myth of the "flesh-eating bananas."

In late 1999, a chain email began to spread across the internet, reporting that fruit imported from Central America could infect people with "necrotising fasciitis"—a rare disease in which the skin erupts into livid purple boils before disintegrating and peeling away from muscle and bone. The email stated that:

> Recently this disease has decimated the monkey population in Costa Rica . . . It is advised not to purchase bananas for the next three weeks as this is the period of time for which bananas that have been shipped to the US with the possibility of carrying this disease. If you have eaten a banana in the last 2–3 days and come down with a fever followed by a skin infection seek MEDICAL ATTENTION!!!
>
> The skin infection from necrotizing fasciitis is very painful and eats two to three centimeters of flesh per hour. Amputation is likely, death is possible. If you are more than an hour from a medical center burning the flesh ahead of the infected area is advised to help slow the spread of the infection. The FDA has been reluctant to issue a country wide warning because of fear of a nationwide panic. They have secretly admitted that they feel upwards of 15,000 Americans will be affected by this but that these are "acceptable numbers." Please forward this to as many of the people you care about as possible as we do not feel 15,000 people is an acceptable number.

By January 28, 2000, public concern was great enough for the US Centers for Disease Control and Prevention to issue a statement denying the risks. But their response only poured fuel on the flames, as people forgot the correction but remembered the scary, vivid idea of the flesh-eating bananas. Some of the chain emails even started citing the CDC as the source of the rumors, giving them greater credibility.

Within weeks, the CDC was hearing from so many distressed callers that it was forced to set up a banana hotline, and it was only by the end of the year that the panic burned itself out as the feared epidemic failed to materialize.[1]

~

The necrotising-fasciitis emails may have been one of the first internet memes—but misinformation is not a new phenomenon. As the eighteenth-century writer Jonathan Swift wrote in an essay on the rapid spread of political lies: "Falsehood flies and the truth comes limping after it."

Today, so-called "fake news" is more prevalent than ever. One survey in 2016 found that more than 50 percent of the most shared medical stories on Facebook had been debunked by doctors, including the claim that "dandelion weed can boost your immune system and cure cancer" and reports that the HPV vaccine increased your risk of developing cancer.[2]

The phenomenon is by no means restricted to the West—though the particular medium may depend on the country. In India, for instance, false rumors spread like wildfire through WhatsApp across its 300 million smartphones—covering everything from local salt shortages to political propaganda and wrongful allegations of mass kidnappings. In 2018, these rumors even triggered a spate of lynchings.[3]

You would hope that traditional education could protect us from these lies. As the great American philosopher John Dewey wrote in the early twentieth century: "If our schools turn out their pupils in that attitude of mind which is conducive to good judgment in any department of affairs in which the pupils are placed, they have done more than if they sent out their pupils merely possessed of vast stores of information, or high degrees of skill in specialized branches."[4]

Unfortunately, the work on dysrationalia shows us this is far from being the case. While university graduates are less likely than average to believe in political conspiracy theories, they are slightly *more* susceptible to misinformation about medicine, believing that pharmaceutical companies are withholding cancer drugs for profit or that doctors are hiding the fact that vaccines *cause* illnesses, for instance.[5] They are also more likely to use unproven, complementary medicines.[6]

It is telling that one of the first people to introduce the flesh-eating banana scare to Canada was Arlette Mendicino, who worked at the University of Ottawa's medical faculty—someone who should have been more skeptical.[7] "I thought about my family, I thought about my friends. I had good intentions," she told CBC News after she found out she'd been fooled. Within a few days, the message had spread across the country.

In our initial discussion of the intelligence trap, we explored the reasons why having a higher IQ might make you ignore contradictory information, so that you are even more tenacious in your existing beliefs, but this didn't really explain why someone like Mendicino could be so gullible in the first place. Clearly this involves yet more reasoning skills that aren't included in the traditional definitions of general intelligence, but that are essential if we want to become immune to these kinds of lies and rumors.

The good news is that certain critical thinking techniques *can* protect us from being duped, but to learn how to apply them, we first need to understand how certain forms of misinformation are designed to escape deliberation and why the traditional attempts to correct them often backfire so spectacularly. This new understanding not only teaches us how to avoid being duped ourselves; it is also changing the way that many global organizations respond to unfounded rumors.

~

Before we continue, first consider the following statements and say which is true and which is false in each pairing:

Bees have learned to tell the difference between
Impressionist and Cubist painters
Bees cannot remember left from right

And

Drinking coffee reduces your risk of diabetes
Cracking your knuckles can cause arthritis

And now consider the following opinions, and say which rings true for you:

Woes unite foes
Strife bonds enemies

And consider which of these online sellers you would shop with:

rifoo73 Average user rating: 3.2
edlokaq8 Average user rating: 3.6

We'll explore your responses in a few pages, but reading the pairs of statements you might have had a hunch that one was truthful or more trustworthy than the other. And the reasons why are helping scientists to understand the concept of "truthiness."

The term was first popularized by the American comedian Stephen Colbert in 2005 to describe the "truth that comes from the gut, not from the book" as a reaction to George W. Bush's decision making and the public perception of his thinking. But it soon became clear that the concept could be applied to many situations[8] and it has now sparked serious scientific research.

Norbert Schwarz and Eryn Newman have led much of this work, and to find out more, I visited them in their lab at the University of Southern California in Los Angeles. Schwarz happens to have been one of the leaders in the new science of emotional decision making that we touched on in the last chapter, showing, for instance, the way the weather sways our judgment of *apparently* objective choices. The work on truthiness extends this idea to examine how we intuitively judge the merits of new information.

According to Schwarz and Newman, truthiness comes from two particular feelings: familiarity (whether we feel that we have heard something like it before) and fluency (how easy a statement is to process). Importantly, most people are not even aware that these two subtle feelings are influencing their judgment—yet they can nevertheless move us to believe a statement without questioning its underlying premises or noting its logical inconsistencies.

As a simple example, consider the following question from some of Schwarz's earlier studies of the subject:

> How many animals of each kind did Moses take on the Ark?

The correct answer is, of course, zero. Moses didn't have an ark—it was Noah who weathered the flood. Yet even when assessing highly intelligent students at a top university, Schwarz has found that just 12 percent of people register that fact.[9]

The problem is that the question's phrasing fits into our basic conceptual understanding of the Bible, meaning we are distracted by the red herring—the quantity of animals—rather than focusing on the name of the person involved. "It's some old guy who had something to do with the Bible, so the whole gist is OK," Schwarz told me. The question turns us into a cognitive miser, in other words—and even the smart university students in Schwarz's study didn't notice the fallacy.

Like many of the feelings fueling our intuitions, fluency and familiarity *can* be accurate signals. It would be too exhausting to examine everything in extreme detail, particularly if it's old news; and if we've heard something a few times, that would suggest that it's a consensus opinion, which may be more likely to be true. Furthermore, things that seem superficially straightforward often are exactly that; there's no hidden motive. So it makes sense to trust things that feel fluent.

What's shocking is just how easy it is to manipulate these two cues with simple changes to presentation so that we miss crucial details.

In one iconic experiment, Schwarz found that people are more likely to fall for the Moses illusion if that statement is written in a pleasant, easy-to-read font—making the reading more fluent—compared to an uglier, italic script that is harder to process. For similar reasons, we are also more likely to believe people talking in a recognizable accent, compared to someone whose speech is harder to understand, and we place our trust in online vendors with easier-to-pronounce names, irrespective of their individual ratings and reviews by other members. Even a simple rhyme can boost the "truthiness" of a statement, since the resonating sounds of the words makes it easier for the brain to process.[10]

Were you influenced by any of these factors in those questions at the start of this chapter? For the record, bees really can be trained to distinguish Impressionist and Cubist painters (and they do also seem to distinguish left from right); coffee can reduce your risk of diabetes, while cracking your knuckles does *not* appear to cause arthritis.[11] But if you are like most people, you may have been swayed by the subtle differences in the way the statements were presented—with the fainter, gray ink and ugly fonts making the true statements harder to read, and less "truthy" as a result. And although they mean exactly the same thing, you are more likely to endorse "woes unite foes" than "strife bonds enemies"—simply because it rhymes.

Sometimes, increasing a statement's truthiness can be as simple as adding an irrelevant picture. In one rather macabre experiment from 2012, Newman showed her participants statements about a series of famous figures—such as a sentence claiming that the indie singer Nick Cave was dead.[12] When the statement was accompanied by a stock photo of the singer, they were more likely to believe that the statement was true, compared to the participants who saw only the plain text.

The photo of Nick Cave could, of course, have been taken at any point in his life. "It makes no sense that someone would use it as evidence—it just shows you that he's a musician in a random band," Newman told me. "But from a psychological perspective it made sense. Anything that would make it easy to picture or easy to imagine something should sway someone's judgment." Newman has also tested the principle on a range of general knowledge statements; they were more likely to agree that "magnesium is the liquid metal inside a thermometer" or "giraffes are the only mammal that cannot jump" if the statement was accompanied by a picture of the thermometer or giraffe. Once again, the photos added no further evidence, but significantly increased the participants' acceptance of the statement.

Interestingly, detailed verbal descriptions (such as of the celebrities' physical characteristics) provided similar benefits. If we are concerned about whether he is alive or dead, it shouldn't matter if we hear that Nick Cave is a white, male singer—but those small, irrelevant details really do make a statement more persuasive.

Perhaps the most powerful strategy to boost a statement's truthiness is simple repetition. In one study, Schwarz's colleagues handed out a list of statements that were said to come from members of the "National Alliance Party of Belgium" (a fictitious group invented for the experiment). But in some of the documents, there appeared to be a glitch in the printing, meaning the same statement from the same person appeared three times. Despite the fact that it was clearly providing no new information, the participants reading the repeated statement were subsequently more likely to believe that it reflected the consensus of the whole group.

Schwarz observed the same effect when his participants read notes about a focus group discussing steps to protect a local park. Some participants read quotes from the same particularly mouthy person who made the same point three times; others read a document in which three different people made the same point, or a document in which three people presented separate arguments. As you might expect, the participants were more likely to be swayed by an argument if they heard it from different people all converging on the same idea. But they were almost as convinced by the argument when it came from a single person, multiple times.[13] "It made hardly any difference," Schwarz said. "You are not tracking who said what."

To make matters worse, the more we see someone, the more familiar they become, and this makes them appear to be more trustworthy.[14] A liar can become an "expert"; a lone voice begins to sound like a chorus, just through repeated exposure.

These strategies have long been known to professional purveyors of misinformation. "The most brilliant propagandist technique will yield no success unless one fundamental principle is borne in mind constantly and with unflagging attention," Adolf Hitler noted in *Mein Kampf*. "It must confine itself to a few points and repeat them over and over."

And they are no less prevalent today. The manufacturers of a quack medicine or a fad diet, for instance, will dress up their claims with reassuringly technical diagrams that add little to their argument—with powerful effect. Indeed, one study found that the mere presence of a brain scan can make pseudo-scientific claims seem more credible—even if the photo is meaningless to the average reader.[15]

The power of repetition, meanwhile, allows a small but vocal minority to persuade the public that their opinion is more popular than it really is. This tactic was regularly employed by tobacco industry lobbyists in the 1960s and 70s. The vice president of the Tobacco Institute, Fred Panzer, admitted as much in an internal memo, describing the industry's "brilliantly conceived strategy" to create "doubt about the health charge without actually denying it," by recruiting scientists to regularly question overwhelming medical opinion.[16]

The same strategies will almost certainly have been at play for many other myths. It is extremely common for media outlets to feature prominent climate change deniers (such as Nigel Lawson in the UK) who have no background in the science but who regularly question the link between human activity and rising sea temperatures. With repetition, their message begins to sound more trustworthy—even though it is only the same small minority repeating the same message. Similarly, you may not remember when you first heard that mobile phones cause cancer and vaccines cause autism, and it's quite possible that you may have even been highly doubtful when you did. But each time you read the headline, the claim gained truthiness, and you became a little less skeptical.

To make matters worse, attempts to debunk these claims often backfire, accidentally spreading the myth. In one experiment, Schwarz showed some undergraduate students a leaflet from the US

Centers for Disease Control, which aimed to debunk some of the myths around vaccinations—such as the commonly held idea that we may become ill after getting the flu shot. Within just thirty minutes, the participants had already started to remember 15 percent of the false claims as facts, and when asked about their intentions to act on the information, they reported that they were less likely to be immunised as a result.[17]

The problem is that the boring details of the correction were quickly forgotten, while the false claims lingered for longer, and became more familiar as a result. By repeating the claim—even to debunk it—you are inadvertently boosting its truthiness. "You're literally turning warnings into recommendations," Schwarz told me.

The CDC observed exactly this when they tried to put the banana hoax to rest. It's little wonder. Their headline: "False Internet report about necrotizing fasciitis associated with bananas" was far less digestible—or "cognitively fluent," in the technical terms—than the vivid (and terrifying) idea of a flesh-eating virus and a government cover-up.

In line with the work on motivated reasoning, our broader worldviews will almost certainly determine how susceptible we are to misinformation—partly because a message that already fits with our existing opinions is processed more fluently and feels more familiar. This may help to explain why more educated people seem particularly susceptible to medical misinformation: it seems that fears about healthcare, in general, are more common among wealthier, more middle-class people, who may also be more likely to have degrees. Conspiracies about doctors—and beliefs in alternative medicine—may naturally fit into that belief system.

The same processes may also explain why politicians' lies continue to spread long after they have been corrected—including Donald Trump's theory that Barack Obama was not born in the United States. As you might expect from the research on motivated reasoning, this was particularly believed by Republicans—but even 14 percent of Democrats held the view as late as 2017.[18]

We can also see this mental inertia in the lingering messages of certain advertising campaigns. Consider the marketing of the mouthwash Listerine. For decades, Listerine's adverts falsely claimed that the mouthwash could soothe sore throats and protect consumers from the common cold. But after a long legal battle in the late 1970s, the Federal Trade Commission forced the company to run adverts correcting the myths. Despite a sixteen-month, $10-million-dollar campaign retracting the statements, the adverts were only marginally effective.[19]

~

This new understanding of misinformation has been the cause for serious soul searching in organizations that are attempting to spread the truth.

In an influential white paper, John Cook, then at the University of Queensland, and Stephan Lewandowsky, then at the University of Western Australia, pointed out that most organizations had operated on the "information deficit model"—assuming that misperceptions come from a lack of knowledge.[20] To counter misinformation on topics such as vaccination, you simply offer the facts and try to make sure that as many people see them as possible.

Our understanding of the intelligence trap shows us that this isn't enough: we simply can't assume that smart, educated people will absorb the facts we are giving them. As Cook and Lewandowsky put it: "It's not just *what* people think that matters, but *how* they think."

Their "debunking handbook" offers some solutions. For one thing, organizations hoping to combat misinformation should ditch the "myth-busting" approach where they emphasise the misconception and then explain the facts. A cursory glance at an NHS web page on vaccines, for instance, lists the ten myths, in bold, right at the top of the page.[21] They are then repeated again, as bold headlines, underneath. According to the latest cognitive science, this kind of approach places too much emphasis on the misinformation itself: the presentation means it is processed more fluently than the facts, and the multiple repetitions simply increase its familiarity. As we have seen, those two feelings—of cognitive fluency and familiarity—contribute to the sense of truthiness, meaning that an anti-vaccination campaigner could hardly have done a better job in reinforcing the view.

Instead, Cook and Lewandowsky argue that any attempt to debunk a misconception should be careful to design the page so that the fact stands out. If possible, you should avoid repeating the myth entirely. When trying to combat fears about vaccines, for instance, you may just decide to focus on the scientifically proven, positive benefits. But if it is necessary to discuss the myths, you can at least make sure that the false statements are less salient than the truth you are trying to convey. It's better to headline your article "Flu vaccines are safe and effective" than "Myth: Vaccines can give you the flu."

Cook and Lewandowsky also point out that many organizations may be too earnest in their presentation of the facts—to the point that they overcomplicate the argument, again reducing the fluency of the message. Instead, they argue that it is best to be selective in the evidence you present: sometimes two facts are more powerful than ten.

For more controversial topics, it is also possible to reduce people's motivated reasoning in the way you frame the issue. If you are trying to discuss the need for companies to pay for the fossil fuels they consume, for example, you are more likely to win over conservative voters by calling it a "carbon offset" rather than a "tax," which is a more loaded term and triggers their political identity.

Although my own browse of various public health websites suggests that many institutions still have a long way to go, there are some signs of movement. In 2017, the World Health Organization announced that they had now adopted these guidelines to deal with the misinformation spread by "anti-vaccination" campaigners.[22]

~

But how can we protect ourselves?

To answer that question, we need to explore another form of metacognition called "cognitive reflection," which, although related to the forms of reflection we examined in the previous chapter, more specifically concerns the ways we respond to factual information, rather than emotional self-awareness.

Cognitive reflection can be measured with a simple test of just three questions, and you can get a flavor of what it involves by considering the following example:

- A bat and a ball cost $1.10 in total. The bat costs $1.00 more than the ball. How much does the ball cost? _____ cents
- In a lake, there is a patch of lily pads. Every day, the patch doubles in size. If it takes 48 days for the patch to cover the entire lake, how long would it take for the patch to cover half of the lake? _____ days
- If it takes 5 machines 5 minutes to make 5 widgets, how long would it take 100 machines to make 100 widgets? ____ minutes?

The math required is not beyond the most elementary education, but the majority of people—even students at Ivy League colleges—only answer between one and two of the three questions correctly.[23] That's because they are designed with misleadingly obvious, but incorrect, answers (in this case, $0.10, 24 days, and 100 minutes). It is only once you challenge those assumptions that you can then come to the correct answer ($0.05, 47 days, and 5 minutes).

This makes it very different from the IQ questions we examined in Chapter 1, which may involve complex calculations, but which do not ask you to question an enticing but incorrect lure. In this way, the Cognitive Reflection Test offers a short and sweet way of measuring how we appraise information and our abilities to override the misleading cues you may face in real life, where problems are ill-defined and messages deceptive.[24]

As you might expect, people who score better on the test are less likely to suffer from various cognitive biases—and sure enough scores on the CRT predict how well people perform on Keith Stanovich's rationality quotient.

In the early 2010s, however, a PhD student called Gordon Pennycook (then at the University of Waterloo) began to explore whether cognitive reflection could also influence our broader beliefs. Someone who stops to challenge their intuitions, and think of alternative possibilities, should be less likely to take evidence at face value, he suspected—making them less vulnerable to misinformation. Sure enough, Pennycook found that people with this more analytical thinking style are less likely to endorse magical thinking and complementary medicine. Further studies have shown that they are also less likely to reject the theory of evolution and to believe 9/11 conspiracy theories.

Crucially, this holds even when you control for other potential factors—like intelligence or education—underlining the fact that it's not just your brainpower that really matters; it's whether or not you use it.[25] "We should distinguish between cognitive ability and cognitive style," Pennycook told me. Or, to put it more bluntly: "If you aren't willing to think, you aren't, practically speaking, intelligent." As we have seen with other measures of thinking and reasoning, we are often fairly bad at guessing where we lie on that spectrum. "People that are actually low in analytic [reflective] thinking believe they are fairly good at it."

Pennycook has since built on those findings, with one study receiving particularly widespread attention, including an Ig Nobel Award for research "that first makes you laugh, then makes you think." The study in question examined the *faux* inspirational, "pseudo-profound bullshit" that people often post on social media. To measure people's credulity, Pennycook asked participants to rate the profundity of various nonsense statements. These included random, made-up combinations of words with vaguely spiritual connotations, such as "Hidden meaning transforms unparalleled abstract beauty." The participants also saw real tweets by Deepak Chopra—a New Age guru and champion of so-called "quantum healing" with more than twenty *New York Times* bestsellers to his name. Chopra's thoughts include: "Attention and intention are the mechanics of manifestation" and "Nature is a self-regulating ecosystem of awareness."

A little like the Moses question, those statements might sound as though they make sense; their buzzwords seem to suggest a kind of warm, inspirational message—until you actually think about their content. Sure enough, the participants with lower CRT scores reported seeing greater meaning in these pseudo-profound statements, compared to people with a more analytical mind-set.[26]

Pennycook has since explored whether this "bullshit receptivity" also leaves us vulnerable to fake news—unfounded claims, often disguised as real news stories, that percolate through social media. Following the discussions of fake news during the 2016 presidential election, he exposed hundreds of participants to a range of headlines—some of which had been independently verified and fact-checked as being true, others as false. The stories were balanced equally between those that were favorable for Democrats and those that were favorable to Republicans.

For example, a headline from the *New York Times* proclaiming that "Donald Trump says he 'absolutely' requires Muslims to register" was supported by a real, substantiated news story. The headline "Mike Pence: Gay conversion therapy saved my marriage" failed fact-checking, and came from the site NCSCOOPER.com.

Crunching the data, Pennycook found that people with greater cognitive reflection were better able to discern the two, regardless of whether they were told the name of the news source, and whether it supported their own political convictions: they were actually engaging with the words themselves and testing whether they were credible rather than simply using them to reinforce their previous prejudices.[27]

Pennycook's research would seem to imply that we could protect ourselves from misinformation by trying to think more reflectively—and a few recent studies demonstrate that even subtle suggestions can have an effect. In 2014, Viren Swami (then at the University of Westminster) asked participants to complete simple word games, some of which happened to revolve around words to do with cognition like "reason," "ponder," and "rational," while others evoked physical concepts like "hammer" or "jump."

After playing the games with the "thinking" words, participants were better at detecting the error in the Moses question, suggesting that they were processing the information more carefully. Intriguingly, they also scored lower on measures of conspiracy theories, suggesting that they were now reflecting more carefully on their existing beliefs, too.[28]

The problems come when we consider how to apply these results to our daily lives. Some of the mindfulness techniques should train you to have a more analytic point of view, and to avoid jumping to quick conclusions about the information you receive.[29] One tantalizing experiment has even revealed that a single meditation can improve scores on the Cognitive Reflection Test, which would seem promising if it can be borne out through future research that specifically examines the effect on the way we process misinformation.[30]

Schwarz is skeptical about whether we can protect ourselves from *all* misinformation through mere intention and goodwill, though: the sheer deluge means that it could be very difficult to apply our skepticism even-handedly. "You couldn't spend all day checking every damn thing you encounter or that is said to you," he told me.*

When it comes to current affairs and politics, for instance, we already have so many assumptions about which news sources are trustworthy—whether it's the *New York Times*, Fox News, Breitbart, or your uncle—and these prejudices can be hard to overcome. In the worst scenario, you may forget to challenge much of the information that agrees with your existing point of view, and only analyze material you already dislike. As a consequence, your well-meaning attempts to protect yourself from bad thinking may fall into the trap of motivated reasoning. "It could just add to the polarization of your views," Schwarz said.

This caution is necessary: we may never be able to build a robust psychological shield against *all* the misinformation in our environment. Even so, there is now some good evidence that we can bolster our defences against the most egregious errors while perhaps also

---

* Pennycook has, incidentally, shown that reflective thinking is negatively correlated with smartphone use—the more you check Facebook, Twitter, and Google, the less well you score on the CRT. He emphasises that we don't know if there is a causal link—or which direction that link would go—but it's possible that technology has made us lazy thinkers. "It might make you more intuitive because you are less used to reflecting—compared to if you are not looking things up, and thinking about things more."

cultivating a more reflective, wiser mind-set overall. We just need to do it more smartly.

Like Patrick Croskerry's attempts to de-bias his medical students, these strategies often come in the form of an "inoculation"—exposing us to one type of bullshit, so that we will be better equipped to spot other forms in the future. The aim is to teach us to identify some of the warning signs, planting little red flags in our thinking, so that we automatically engage our analytical, reflective reasoning when we need it.

John Cook and Stephan Lewandowsky's work suggests the approach can be very powerful. In 2017, Lewandowsky and Cook (who also wrote *The Debunking Handbook*) were investigating ways to combat some of the misinformation around human-made climate change—particularly the attempts to spread doubt about the scientific consensus.

Rather than tackling climate myths directly, however, they first presented some of their participants with a fact sheet about the way the tobacco industry had used "fake experts" to cast doubts on scientific research linking smoking to lung cancer.

They then showed them a specific piece of misinformation about climate change: the so-called Oregon Petition, organized by the biochemist Arthur B. Robinson, which claimed to offer 31,000 signatures of people with science degrees, who all doubted that human release of greenhouse gases is causing disruption of the Earth's climate. In reality, the names were unverified—the list even included the signature of Spice Girl "Dr" Geri Halliwell[31]—and fewer than 1 percent of those questioned had formally studied climate science.

Previous research had shown that many people reading about the petition fail to question the credentials of the experts, and are convinced of its findings. In line with theories of motivated reasoning, this was particularly true for people who held more right-wing views.

After learning about the tobacco industry's tactics, however, most of Cook's participants were more skeptical of the misinformation, and it failed to sway their overall opinions. Even more importantly, the inoculation had neutralized the effect of the misinformation *across the political spectrum*; the motivated reasoning that so often causes us to accept a lie, and reject the truth, was no longer playing a role.[32] "For me that's the most interesting result—inoculation works despite your political background," Cook told me. "Regardless of ideology, no one wants to be misled by logical fallacies—and that is an encouraging and exciting thought."

Equally exciting is the fact that the inoculation concerning misinformation in one area (the link between cigarettes and smoking) provided protection in another (climate change). It was as if participants had planted little alarm bells in their thinking, helping them to recognize when to wake up and apply their analytic minds more effectively, rather than simply accepting any information that felt "truthy." "It creates an umbrella of protection."

~

The power of these inoculations is leading some schools and universities to explore the benefits of explicitly educating students about misinformation.[33]

Many institutions already offer critical thinking classes, of course, but these are often dry examinations of philosophical and logical principles, whereas inoculation theory shows that we need to be taught about it explicitly, using real-life examples that demonstrate the kinds of arguments that normally fool us.[34] It does not seem to be enough to assume that we will readily apply those critical thinking skills in our everyday lives without first being shown the sheer prevalence of misinformation and the ways it could be swaying our judgments.

The results so far have been encouraging, showing that a semester's course in inoculation significantly reduced the students' beliefs in pseudoscience, conspiracy theories, and fake news. Even more importantly, these courses also seem to improve measures of critical thinking more generally—such as the ability to interpret statistics, identify logical fallacies, consider alternative explanations, and recognize when additional information will be necessary to come to a conclusion.[35]

Although these measures of critical thinking are not identical to the wise reasoning tests we explored in Chapter 5, they do bear some similarities—including the ability to question your own assumptions and to explore alternative explanations for events. Importantly, like Igor Grossmann's work on evidence-based wisdom, and the scores of emotion differentiation and regulation that we explored in the last chapter, these measures of critical thinking don't correlate very strongly with general intelligence, and they predict real-life outcomes better than standard intelligence tests.[36] People with higher scores are less likely to try an unproven fad diet, for instance; they are also less likely to share personal information with a stranger online or to have unprotected sex. If we are smart but want to avoid making stupid mistakes, it is therefore essential that we learn to think more critically.

These results should be good news for readers of this book: by studying the psychology of these various myths and misconceptions, you may have already begun to protect yourself from lies—and the existing cognitive inoculation programs already offer some further tips to get you started.

The first step is to learn to ask the right questions:

- Who is making the claim? What are their credentials? And what might be their motives to make me think this?
- What are the premises of the claim? And how might they be flawed?
- What are my own initial assumptions? And how might they be flawed?
- What are the alternative explanations for their claim?
- What is the evidence? And how does it compare to the alternative explanation?
- What further information do you need before you can make a judgment?

Given the research on truthiness, you should also look at the presentation of the claims. Do they actually add any further proof to the claim—or do they just give the illusion of evidence? Is the same person simply repeating the same point—or are you really hearing different voices who have converged on the same view? Are the anecdotes offering useful information and are they backed up with hard data? Or do they just increase the fluency of the story? And do you trust someone simply because their accent feels familiar and is easy to understand?

Finally, you should consider reading about a few of the more common logical fallacies, since this can plant those "red flags" that will alert you when you are being duped by "truthy" but deceptive information. To get you started, I've compiled a list of the most common ones in the table below.

These simple steps may appear to be stating the obvious, but overwhelming evidence shows that many people pass through university without learning to apply them to their daily life.[37] And the overconfidence bias shows that it's the people who think they are already immune who are probably most at risk.

Fallacy	Explanation	Example
Appeal to ignorance	A lack of evidence is taken as a form of proof.	Our inability to explain how the Egyptians made the pyramids means that they must have been built by aliens.
Appeals to authority	Considering that someone's credentials proves that they *must* be right, even if that contradicts other evidence. This is problematic if the expert's opinion is controversial within their own field.	Kary Mullis, a Nobel Prize–winning biologist, claims that HIV does not lead to AIDS—and if someone so intelligent thinks this, then *it must be true.* You can also see this fallacy when athletes endorse health supplements. Just because they have excellent fitness doesn't mean their nutritional advice is valid.
Correlation proves causation	When two events coincide, we believe that one led to the other, without considering other factors.	People eating a fad diet may live longer, but that could equally be caused by the fact that they are more fitness conscious and also do more exercise.

Straw man arguments	Deliberately misrepresenting an argument to make it seem more ludicrous. In conversation, this may come in the form, "So what you're saying, is . . ." followed by an incorrect or oversimplified summary.	Darwin's theories were used to justify racial differences; therefore the theory of evolution is itself a racist ideology. (This was an actual argument under consideration by the Louisiana State Legislature, when it was reviewing its educational policy.)[38] Incidentally, similar arguments are often used to dismiss IQ—but even if Lewis Terman's political beliefs were questionable, that should not be used to judge the scientific results.
Appeal to the bandwagon	The idea that popular opinion proves an argument's value.	Millions of people anecdotally claim that homeopathy improves their symptoms; therefore it must be a valid treatment.
False dichotomy	Presenting a complex scenario as if there are only two options when there are many other options.	"Every nation, in every region, now has a decision to make. Either you are with us, or you are with the terrorists"—George Bush following 9/11.
Red herrings	Using irrelevant information to distract people from the flaws in the actual argument.	"Secondhand smoking may be dangerous; but people will always overeat and drink too much alcohol even if we ban cigarettes." The second point, of course, is irrelevant to the first—but it is presented as if it provides further evidence.
Special pleading	Claiming that the normal rules of logic and evidence cannot apply to the case in question.	Psychics often claim that scientific experiments (and the scientists' skepticism) interfere with their abilities. Arthur Conan Doyle was particularly guilty of this fallacy.

If you really want to protect yourself from bullshit, I can't overemphasise the importance of internalizing these rules and applying them whenever you can, to your own beloved theories as well as those that already arouse your suspicion. If you find the process rewarding, there are plenty of online courses that will help you to develop those skills further.

According to the principles of inoculation, you should start out by looking at relatively uncontroversial issues (like the flesh-eating bananas) to learn the basics of skeptical thought, before moving on to more deeply embedded beliefs (like climate change) that may be harder for you to question. In these cases, it is always worth asking why you feel strongly about a particular viewpoint, and whether it is really central to your identity, or whether you might be able to reframe it in a way that is less threatening.

Simply spending a few minutes to write positive, self-affirming things about yourself and the things that you most value can make you more open to new ideas. Studies have shown that this practice really does reduce motivated reasoning by helping you to realize that your whole being does not depend on being right about a particular issue, and that you can disentangle certain opinions from your identity.[39] (Belief in climate change does not have to tear down your conservative politics, for instance: you could even see it as an opportunity to further business and innovation.) You can then begin to examine why you might have come to those conclusions, and to look at the information in front of you and test whether you might be swayed by its fluency and familiarity.

You may be surprised by what you find. Applying these strategies, I've already changed my mind on certain scientific issues, such as genetic modification. Like many liberal people, I had once opposed GM crops on environmental grounds—yet the more I became aware of my news sources, the more I noticed that I was hearing opposition from the same small number of campaign groups like Greenpeace—creating the impression that these fears were more widespread than they actually were. Moreover, their warnings about toxic side effects and runaway plagues of Frankenstein plants were cognitively fluent and chimed with my intuitive environmental views—but a closer look at the evidence showed that the risks are tiny (and mostly based on anecdotal data), while the potential benefits of building insect-resistant crops and reducing the use of pesticides are incalculable.

Even the former leader of Greenpeace has recently attacked the scaremongering of his ex-colleagues, describing it as "morally unacceptable . . . putting ideology before humanitarian action."[40] I had always felt scornful of climate change deniers and anti-vaccination campaigners, yet I had been just as blinkered concerning another cause.

~

For one final lesson in the art of bullshit detection, I met the writer Michael Shermer in his home town of Santa Barbara, California. For the past three decades, Shermer has been one of the leading voices of the skeptical movement, which aims to encourage the use of rational reasoning and critical thinking to public life. "We initially went for the low-hanging fruit—television psychics, astrology, tarot card reading," Shermer told me. "But over the decades we've migrated to more 'mainstream' claims about things like global warming, creationism, anti-vaccination—and now fake news."

Shermer has not always been this way. A competitive cyclist, he once turned to unproven (though legal) treatments to boost his performance, including colonic irrigation to ease his digestion, and "Rolfing"—a kind of intense (and painful) physiotherapy which involves manipulating the body's connective tissue to reinforce its "energy field." At night, he had even donned an "Electro-Acuscope"—a device, worn over the skull, that was designed to enhance the brain's healing "alpha waves."

Shermer's "road-to-Damascus moment" came during the 1983 Race Across America, from Santa Monica, California, to Atlantic City, New Jersey. For this race, Shermer hired a nutritionist, who advised him to try a new "multivitamin therapy"—which involved ingesting a mouthful of foul-smelling tablets. The end result was the "most expensive and colorful urine in America." By the third day, he decided that enough was enough—and on the steep climb to Loveland Pass, Colorado, he spat out the mouthful of acrid tablets and vowed never to be duped again. "Being skeptical seemed a lot safer than being credulous," he later wrote.[41]

A stark test of his newfound skepticism came a few days later, near Haigler, Nebraska. It was nearly halfway through the race and he was already suffering from severe exhaustion. After waking from a forty-five-minute nap, he was convinced that he was surrounded by aliens, posing as his crew members, trying to take him to the mothership. He fell back asleep, and awoke clear-headed and realized that he had experienced a hallucination arising from physical and mental exhaustion. The memory remains vivid, however, as if it were a real event. Shermer thinks that if he had not been more self-aware, he could have genuinely confused the event for a real abduction, as many others before him have done.

As a historian of science, writer, and public speaker, Shermer has since tackled psychics, quack doctors, 9/11 conspiracy theorists, and holocaust deniers. He has seen how your intelligence can be applied powerfully to either discover or obfuscate the truth.

You might imagine that he would be world-weary and cynical after so many years of debunking bullshit, yet he was remarkably affable on our meeting. A genial attitude is, I later found out, crucial for putting many of his opponents off their guard, so that he can begin to understand what motivates them. "I might socialize with someone like [Holocaust denier] David Irving, because after a couple of drinks, they open up and go deeper, and tell you what they are really thinking."[42]

Shermer may not use the term, but he now offers one of the most comprehensive "inoculations" available in his "Skepticism 101" course at Chapman University.[43] The first steps, he says, are like "kicking the tires and checking under the hood" of a car. "Who's making the claim? What's the source? Has someone else verified the claim? What's the evidence? How good is the evidence? Has someone tried to debunk the evidence?" he told me. "It's basic baloney detection."

Like the other psychologists I have spoken to, he is certain that the vivid, real-life examples of misinformation are crucial to teach these principles; it's not enough to assume that a typical academic education equips us with the necessary protection. "Most education is involved in just teaching students facts and theories about a particular field—not necessarily the methodologies of thinking skeptically or scientifically in general."

To give me a flavor of the course, Shermer describes how many conspiracy theories use the "anomalies-as-proof" strategy to build a superficially convincing case that something is amiss. Holocaust deniers, for instance, argue that the structure of the (badly damaged) Krema II gas chamber at Auschwitz-Birkenau doesn't match eyewitness accounts of SS guards dropping gas pellets through the holes in the roof. From this, they claim that no one could have been gassed at Krema II, therefore no one would have been gassed at Auschwitz-Birkenau, meaning that no Jews were systematically killed by the Nazis—and the Holocaust didn't happen.

If that kind of argument is presented fluently, it may bypass our analytical thinking; never mind the vast body of evidence that does not hinge on the existence of holes in Krema II, including aerial photographs showing mass exterminations, the millions of skeletons in mass graves, and the confessions of many Nazis themselves. Attempts to reconstruct the Krema gas chamber have, in fact, found the presence of these holes, meaning the argument is built on a false premise—but the point is that even if the anomaly had been true, it wouldn't have been enough to rewrite the whole of Holocaust history.

The same strategy is often used by people who believe that the 9/11 attacks were "an inside job." One of their central claims is that jet fuel from the airplanes could not have burned hot enough to melt the steel girders in the Twin Towers, meaning the buildings should not have collapsed. (Steel melts at around 2,750° F, whereas the fuel from the airplanes burns at around 1,520° F.) In fact, although steel does not turn into a liquid at that temperature, engineers have shown that it nevertheless loses much of its strength, meaning the girders would have nevertheless buckled under the weight of the building. The lesson, then, is to beware of the use of anomalies to cast doubt on vast sets of data, and to consider the alternative explanations before you allow one puzzling detail to rewrite history.[44]

Shermer emphasises the importance of keeping an open mind. With the Holocaust, for instance, it's important to accept that there will be some revising of the original accounts as more evidence comes to light, without discounting the vast substance of the accepted events.

He also advises us all to step outside of our echo chamber and to use the opportunity to probe someone's broader worldviews; when talking to a climate change denier, for instance, he thinks it can be useful to explore their economic concerns about regulating fossil fuel consumption—teasing out the assumptions that are shaping their interpretation of the science. "Because the facts about global warming are not political—they are what they are." These are the same principles we are hearing again and again: to explore, listen, and learn, to look for alternative explanations and viewpoints rather than the one that comes most easily to mind, and to accept you do not have all the answers.

By teaching his students this kind of approach, Shermer hopes that they will be able to maintain an open-minded outlook, while being more analytical about any source of new information. "It's equipping them for the future, when they encounter some claim twenty years from now that I can't even imagine, so they can think, well this is kind of like that thing we learned in Shermer's class," he told me. "It's just a toolkit for anyone to use, at any time. . . . This is what all schools should be doing."

~

Having first explored the foundations of the intelligence trap in Part 1, we've now seen how the new field of evidence-based wisdom outlines additional thinking skills and dispositions—such as intellectual humility, actively open-minded thinking, emotion differentiation and regulation, and cognitive reflection—and helps us to take control of the mind's powerful thinking engine, circumventing the pitfalls that typically afflict intelligent and educated people.

We've also explored some practical strategies that allow you to improve your decision making. These include Benjamin Franklin's moral algebra, self-distancing, mindfulness, and reflective reasoning, as well as various techniques to increase your emotional self-awareness and fine-tune your intuition. And in this chapter, we have seen how these methods, combined with advanced critical thinking skills, can protect us from misinformation: they show us to beware of the trap of cognitive fluency, and they should help us to build wiser opinions on politics, health, the environment, and business.

One common theme is the idea that the intelligence trap arises because we find it hard to pause and think beyond the ideas and feelings that are most readily accessible, and to take a step into a different vision of the world around us; it is often a failure of the imagination at a very basic level. These techniques teach us how to avoid that path, and as Silvia Mamede has shown, even a simple pause in our thinking can have a powerful effect.

Even more important than the particular strategies, however, these results are an invaluable proof of concept. They show that there are indeed many vital thinking skills, besides those that are measured in standard academic tests, that can guide your intelligence to ensure that you use it with greater precision and accuracy. And although these skills are not currently cultivated within a standard education, they *can* be taught. We can all train ourselves to think more wisely.

In Part 3, we will expand on this idea to explore the ways that evidence-based wisdom can also boost the ways we learn and remember—firmly putting to rest the idea that the cultivation of these qualities will come at the cost of more traditional measures of intelligence. And for that, we first need to meet one of the world's most curious men.

PART 3

The art of successful learning: *How evidence-based wisdom can improve your memory*

7

Tortoises and hares: *Why smart people fail to learn*

Let's return to the US in the late 1920s. In California, Lewis Terman's geniuses have just started to attend high school, the vision of a glittering future still stretching out before them, but we are more interested in a young boy called Ritty, tinkering away in his home laboratory in Far Rockaway, New York.

The "lab" comprised an old wooden packing box, equipped with shelves, a heater, a storage battery, and an electric circuit of light bulbs, switches, and resistors. One of Ritty's proudest projects was a homemade burglar alarm, so that a bell would sound whenever his parents entered his room. He used a microscope to study the natural world and he would sometimes take his chemistry set into the street to perform shows for the other children.

The experiments did not always end as he had planned. One day, he began to play with the ignition coil from a Ford car. Could the sparks punch holes through a piece of paper, he wondered? They did, but before he knew it the paper was ablaze. When it became too hot to hold, Ritty dropped it into a wastepaper bin—which itself caught light. Conscious of his mother playing bridge downstairs, he carefully closed the door, and smothered the fire with an old magazine, before shaking the embers onto the street below.[1]

None of this necessarily marks Ritty as anything out of the ordinary: myriad children of his generation will have owned chemistry sets, played with electric circuits, and studied the natural world with a microscope. He was, by his own admission, a "goody-goody" at school, but by no means remarkable: he struggled with literature, drawing, and foreign languages. Perhaps because of his poorer verbal skills, he apparently scored 125 in a school IQ test, which is above average but nowhere near the level of the "geniuses" in California.[2] Lewis Terman would not have given him much thought compared to the likes of Beatrice Carter, with her astronomical score of 192.

But Ritty kept learning anyway. He devoured the family encyclopedia, and as a young adolescent he soon took to teaching himself from a series of mathematics primers—filling his notebooks with trigonometry, calculus, and analytic geometry, often creating his own exercises to stretch his mind.[3] When he moved to the Far Rockaway High School, he joined a physics club and entered the Interscholastic Algebra League. He eventually reached the top place in New York University's annual math championship—ahead of students from all the city's schools. The next year, he began his degree at MIT—and the rest is history.

Schoolchildren would later learn Ritty's full name—Richard Feynman—as one of the most influential physicists of the twentieth century. His new approach to the field of quantum electrodynamics revolutionized the study of subatomic particles[4]—research that won him a Nobel Prize in 1965 with Sin-Itiro Tomonaga and Julian Schwinger.[5] (It was an accolade that none of Terman's cohort would achieve.) Feynman also helped uncover the physics behind radioactive decay, and made vital contributions to America's development of the atomic bomb during the Second World War, a role that he later deeply regretted.

Other scientists believed that the depths of his thinking were almost unfathomable. "There are two kinds of geniuses: the 'ordinary' and the 'magicians,'" the Polish mathematician Mark Kac wrote in his autobiography. "An ordinary genius is a fellow that you and I would be just as good as, if we were only many times better. There is no mystery as to how his mind works. Once we understand what they have done, we feel certain that we, too, could have done it. It is different with magicians . . . the working of their minds is for all intents and purposes incomprehensible. Even after we understand what they have done, the process by which they have done it is completely dark. . . . Richard Feynman is a magician of the highest calibre."[6]

But Feynman's genius did not end with physics. During a sabbatical from his physics research at Caltech, he applied himself to the study of genetics, discovering the ways that some mutations within a gene may suppress each other. Despite his apparent inaptitude for drawing and foreign languages, he later learned to be a credible artist, to speak Portuguese and Japanese, and to read Mayan hieroglyphs—all with the relentlessness that had driven his education as a child. Other projects included a study of ant behavior, bongo drumming, and a long-standing fascination with radio repair. After the 1986 *Challenger* disaster, it was Feynman's tenacious inquiring mind that exposed the engineering flaw that had caused the space shuttle to explode.

As Feynman's biographer James Gleick wrote in a *New York Times* obituary: "He was never content with what he knew, or what other people knew. . . . He pursued knowledge without prejudice."[7]

~

The stories of Lewis Terman's "geniuses" have already shown us how people of great general intelligence often fail to build on their initial potential. Despite their early promise, many of the Termites reached old age with the uneasy feeling that they could have done more with their talents. Like the hare in Aesop's most famous fable, they began with a natural advantage but failed to capitalize on that potential.

Feynman, in contrast, claimed to have started out with a "limited intelligence,"[8] but he then applied it in the most productive way possible, as he continued to grow and expand his mind throughout adulthood. "The real fun of life," he wrote to a fan in 1986, just two years before he died, "is this perpetual testing to realize how far out you can go with any potentialities."[9]

The latest psychological research on learning and personal development has now started to see an astonishing convergence with the theory of evidence-based wisdom that we have explored so far in this book, revealing additional cognitive qualities and mental habits, besides intelligence, that may determine whether or not we flourish like Feynman.

By encouraging us to engage and stretch our minds, these characteristics can boost our learning and ensure that we thrive when we face new challenges, ensuring that we make the most of our natural potential. Crucially, however, they also provide an antidote to the cognitive miserliness and one-sided thinking that contributes to some forms of the intelligence trap—meaning that they also result in wiser, less biased reasoning overall.

These insights may be of particular interest to parents and people working in education, but they can also empower anyone to apply their intelligence more effectively.

~

Let's first consider curiosity, a trait that appears common in many other high achievers besides Feynman.

Charles Darwin, for instance, had failed to excel in his early education and, like Feynman, he certainly didn't consider himself to be of above average intelligence, claiming that he had "no great quickness of apprehension or wit which is so remarkable in some clever men."[10]

"When I left the school I was for my age neither high nor low in it," he wrote in an autobiographical essay.

> And I believe that I was considered by all my masters and by my father as a very ordinary boy, rather below the common standard in intellect. . . . Looking back as well as I can at my character during my school life, the only qualities which at this period promised well for the future were that I had strong and diversified tastes, much zeal for whatever interested me, and a keen pleasure in understanding any complex subject or thing.[11]

It is difficult to imagine that Darwin could have ever conducted his painstaking work on the *Beagle*—and during the years afterwards—if he had not been driven by a hunger for knowledge and understanding. He certainly wasn't looking for immediate riches or fame: the research took decades with little payoff. But his desire to learn more caused him to look further and question the dogma around him.

Besides his groundbreaking work on evolution, Darwin's ceaseless interest in the world around him would lead to some of the first scientific writings on the subject of curiosity, too, describing how young children naturally learn about the world about them through tireless experimentation.[12]

As later child psychologists noted, this "need to know more" was almost like a basic biological drive, or hunger, for a young infant. Despite this scientific pedigree, however, modern psychologists had largely neglected to systematically explore its broader role in our later lives, or the reasons that some people are naturally more curious than others.[13] We knew that curiosity was crucial for taking our first intellectual steps in the world—but little after that.

That was partly due to practical difficulties. Unlike general intelligence, there are no definitive standardized tests, meaning that psychologists have instead relied on more tangential indicators. You can observe how often a child asks questions, for instance, or how intensely they explore their environment; you can also design toys with hidden features and puzzles, and measure how long the child engages with them. With adults, meanwhile, one can use self-reported questionnaires, or behavioral tests that examine whether someone will read and probe new material or if they are happy to ignore it. And when modern psychologists have turned to these tools, they have found that curiosity can rival general intelligence in its importance over our development throughout childhood, adolescence, and beyond.

Much of that research on curiosity had examined its role in memory and learning,[14] showing that someone's curiosity can determine the amount of material that is remembered, the depth of the understanding, and the length of time that the material is retained.[15] This isn't just a question of motivation: even when their additional effort and enthusiasm is taken into consideration, people with greater curiosity still appear to be able to remember facts more easily.

Brain scans can now tell us why this is, revealing that curiosity activates a network of regions known as the "dopaminergic system." The neurotransmitter dopamine is usually implicated in desire for food, drugs, or sex—suggesting that, at a neural level, curiosity really is a form of hunger or lust. But the neurotransmitter also appears to strengthen the long-term storage of memories in the hippocampus, neatly explaining why curious people are not only more motivated to learn, but will also remember more, even when you account for the amount of work they have devoted to a subject.[16]

The most interesting discovery has been the observation of a "spillover effect"—meaning that once the participants' interest has been piqued by something that genuinely interests them, and they have received that shot of dopamine, they subsequently find it easier to memorize incidental information too. It primes the brain for learning anything.

Importantly, the research shows that some people are consistently more interested in the world around them. And these individual differences in curiosity are only modestly related to general intelligence. This means that two people of the same IQ may have radically different trajectories depending solely on their curiosity, and a genuine interest in the material will be more important than a determination to succeed.

For this reason, some psychologists now consider that general intelligence, curiosity and conscientiousness are together the "three pillars" of academic success; if you lack any one of these qualities, you are going to suffer.

The benefits do not end with education. At work, curiosity is crucial for us to pick up the "tacit knowledge" that we explored in Chapter 1, and it can protect us from stress and burnout, helping us to remain motivated even when the going gets tough. It also powers our creative intelligence, by encouraging us to probe problems that others had not even considered, and by triggering counterfactual thinking as we ask ourselves "what if . . . ?"[17]

A genuine interest in the other person's needs even improves our social skills and helps us to uncover the best potential compromise—boosting our emotional intelligence.[18] By encouraging us to look more deeply for unspoken motivations in this way, curiosity seems to lead to better business negotiations.

The result is a richer and more fulfilling life. One landmark study tracked the lives of nearly eight hundred people over the course of two six-month periods, questioning them about their personal goals. Using self-reported questionnaires to measure ten separate traits—including self-control and engagement—the researchers found that curiosity best predicted their ability to achieve those goals.[19]

If you are wondering how you would compare to these participants, consider the following sample questions and score how accurately they reflect the way you feel and behave, from 1 (not at all) to 5 (extremely):

- I actively seek as much new information as I can in new situations.
- Everywhere I go, I am out looking for new things or experiences.
- I am the kind of person who embraces unfamiliar people, events and places.[20]

The people who strongly endorsed these kinds of statements were more likely to succeed at whatever they set their mind to achieve. Curiosity was also the only trait that consistently boosted well-being during those twelve months. In other words, it didn't just increase their chances of success; it made sure that they enjoyed the process too.

All of which helps us to understand how people like Darwin and Feynman could achieve so much in their lives. The hunger to explore had exposed them to new experiences and ideas that didn't fit with the current orthodoxy; it then drove them to dig deeper to understand what they were seeing and to find novel solutions to the problems they uncovered.

Someone with greater intelligence might have initially found it easier to process complex information than either of these two men, but if they lacked a natural curiosity they are unlikely to have been able to maintain that advantage. It shows us, again, that general intelligence is one crucial ingredient of good thinking—but it needs many other complementary traits to truly flourish.

The real mystery is why so few of us manage to maintain that childlike interest, with many studies showing that most people's curiosity drops rapidly after infancy. If we are all born with a natural hunger to learn, and that trait can bring us so many benefits well into adulthood, what causes so many people to lose it as we age? And how can we stop that decline?

Susan Engel, at Williams College, Massachusetts, has spent the best part of the last two decades looking for answers—and the results are shocking. In her book *The Hungry Mind*, she points to one experiment, in which a group of kindergarten children were allowed to watch one of their parents in a separate room through one-way glass. The parents were either asked to play with the objects on a table, to simply look at the table, or to ignore the objects completely as they chatted to another adult. Later on, the children were given the objects to inspect—and they were far more likely to touch and explore them if they had seen their parents doing the same.

Through the subtlest of actions, their parents' behavior had shown the children whether exploration was desired or discouraged, enhancing or damping their interest, and over time, these attitudes could become ingrained in their minds. "Curiosity is contagious, and it's very difficult to encourage curiosity in kids if you don't have any experience of curiosity in your own life," Engel said.

A parent's influence also comes through their conversation. Recording twelve families' dinner-table conversations, she noticed that some parents routinely offer a straight answer to a child's questions. There was nothing actually wrong with what they said—they were not notably uninterested—but others used the opportunity to open up the subject, which would inevitably lead to a chain of further questions. The result was a far more curious and engaged child.

Engel's research paints an even bleaker picture of our education systems. Toddlers may ask up to twenty-six questions per hour at home (with one child asking 145 during one observation!) but this drops to just two per hour at school. This disengagement can also be seen in other expressions of curiosity—such as how willing they are to explore new toys or interesting objects—and it becomes even more pronounced as the child ages. While observing some fifth-grade lessons, Engel would often go for a two-hour stretch without seeing a single expression of active interest.

This may partly be due to teachers' understandable concerns about maintaining order and meeting the demands of their syllabus. Even so, Engel believes that many teachers are often too rigid, failing to let students pursue their own questions in favor of adhering to a predefined lesson plan. When observing one class on the American Revolution, for instance, she saw one boy politely raise his hand after fifteen minutes of nonstop lecture. "I can't answer questions right now," the teacher replied in a brisk tone. "Now it's time for learning." You can see how that attitude could quickly rub off on a child, so that even someone of greater intelligence simply stops trying to find things out for themselves.

Darwin, incidentally, had found that rigid classical education almost killed his interest, as he was forced to learn Virgil and Homer by heart. "Nothing could have been worse for the development of my mind," he wrote. Fortunately, he had at least been encouraged to pursue his interests by his parents. But without any nourishment at home or at school, your appetite to learn and explore may slowly disappear.

Engel points out that anxiety is also a curiosity killer, and very subtle cues may have a big impact; she has even found that the expression of interest is directly correlated with the number of times a teacher smiles during the lesson.

In another experiment, she studied groups of nine-year-olds in a science lesson. Their task was simple—they had to drop raisins into a mixture of vinegar, baking soda, and water, to see if the bubbles would make them float. In half of the lessons, the teacher set out the instructions and then left the children to get on with their work, but in others, the teacher deviated from the lesson plan slightly. She picked up a Skittle candy and said, "You know what, I wonder what would happen if I dropped this instead."

It was a tiny step, but having observed the teacher's expression of curiosity, the children engaged more enthusiastically with the lesson—continuing their endeavours even when the teacher left the room. It was a stark contrast to the control condition, in which the children were more easily distracted, more fidgety, and less productive.

Although Engel's work is still ongoing, she is adamant that it's time to bring these insights into the classroom. "There's a lot we still don't know and that's a very exciting thing for us as scientists. But we know enough to say that schools should be [actively] encouraging curiosity . . . and that it can be very powerful. A kid who really wants to know things—you practically can't stop them from learning."

~

We will soon discover the ways that Feynman managed to keep his curiosity to achieve his potential—and why this also contributes to better reasoning and thinking. Before we examine those groundbreaking discoveries, however, we also need to examine one other essential ingredient for personal and intellectual fulfillment: a characteristic known as the "growth mind-set."

This concept is the brainchild of Carol Dweck, a psychologist at Stanford University, whose pioneering research first attracted widespread attention in 2007 with a best-selling book, *Mindset*. But this was just the beginning. Over the last decade, a series of striking experiments has suggested that our mind-sets can also explain why apparently smart people fail to learn from their errors, meaning that Dweck's theory is essential for our understanding of the intelligence trap.

Like Robert Sternberg, Dweck was inspired by her own experience at school. During sixth grade, Dweck's teacher seated the class according to IQ—with the "best" at the front, and the "worst" at the back. Those with the lowest scores were not even allowed the menial tasks of carrying the flag or taking a note to the principal. Although she was placed in row one, seat one, Dweck felt the strain of the teacher's expectations.[21] "She let it be known that IQ for her was the ultimate measure of your intelligence and your character."[22] Dweck felt that she could trip up at any moment, which made her scared to try new challenges.

She would remember those feelings when she started her work as a developmental psychologist. She began with a group of ten- and eleven-year-olds, setting them a number of stretching logical puzzles. The children's success at the puzzles was not necessarily linked to their talent; some of the brightest quickly became frustrated and gave up, while others persevered.

The difference instead seemed to lie in their beliefs about their own talents. Those with the growth mind-set had faith that their performance would improve with practice, while those with the fixed mind-set believed that their talent was innate and could not be changed. The result was that they often fell apart with the more challenging problems, believing that if they failed now, they would fail for ever. "For some people, failure is the end of the world, but for others, it's an exciting new opportunity."[23]

In experiments across schools, universities, and businesses, Dweck has now identified many attitudes that might cause smart people to develop the fixed mind-set. Do you, for instance, believe that:

- A failure to perform well at the task at hand will reflect your overall self-worth?
- Learning a new, unfamiliar task puts you at risk of embarrassment?
- Effort is only for the incompetent?
- You are too smart to try hard?

If you broadly agree with these statements, then you may have more of a fixed mind-set, and you may be at risk of sabotaging your own chances of later success by deliberately avoiding new challenges that would allow you to stretch beyond your comfort zone.[24]

At Hong Kong University, for instance, Dweck measured the mind-set of students entering their first year on campus. All the lessons are taught in English, so proficiency in that language is vital for success, but many students had grown up speaking Cantonese at home and were not perfectly fluent. Dweck found that students with the fixed mind-set were less enthusiastic about the possibility of taking an English course, as they were afraid it might expose their weakness, even though it could increase their long-term chances of success.[25]

Besides determining how you respond to challenge and failure, your mind-set also seems to influence your ability to learn from the errors you do make—a difference that shows up in the brain's electrical activity, measured through electrodes placed on the scalp. When given negative feedback, people with the fixed mind-set show a heightened response in the anterior frontal lobe—an area known to be important for social and emotional processing, with the neural activity appearing to reflect their bruised egos. Despite these strong emotions, however, they showed less activity in the temporal lobe, associated with deeper conceptual processing of the information. Presumably, they were so focused on their hurt feelings, they weren't concentrating on the details of what was actually being said and the ways it might improve their performance next time. As a result, someone with the fixed mind-set is at risk of making the same mistakes again and again, leading their talents to founder rather than flourish.[26]

In school, the consequences may be particularly important for children from less advantaged backgrounds. In 2016, for instance, Dweck's team published the result of a questionnaire that examined the mind-sets of more than 160,000 tenth-graders in Chile—the first sample across a whole nation. As the previous research would have predicted, a growth mind-set predicted academic success across the group, but the team also examined the way it benefited the less privileged children in the group. Although the poorest 10 percent of children were more likely to have a fixed mind-set, the researchers found those with the growth mind-set tended to perform as well as the richest children in the sample, from families who earned thirteen times more money. Although we can only read so much from a correlational study, the growth mind-set seemed to be driving them to overcome the many hurdles associated with their poverty.[27]

Beyond education, Dweck has also worked with racing car drivers, professional football players, and Olympic swimmers to try to change their mind-sets and boost their performance.[28]

Even people at the height of their career can find themselves constrained by a fixed mind-set. Consider the tennis player Martina Navratilova, the world champion who lost to the sixteen-year-old Italian Gabriela Sabatini at the Italian Open in 1987. "I felt so threatened by those younger kids," she later said, "I daren't give those matches 100%. I was scared to find out if they could beat me when I'm playing my best."[29]

Navratilova identified and adjusted this outlook, and went on to win Wimbledon and the US Open, but some people may spend their whole lives avoiding challenge. "I think that's how people live narrow lives," Dweck told me. "You take this chance to play it safe, but if you add up all those moments you are far into the future and you haven't expanded yourself."

Dweck's research has gained widespread acclaim, but the attention is not always well directed, with many people misreading and misinterpreting her work. A *Guardian* article from 2016, for instance, described it as "the theory that anyone who tries can succeed,"[30] which isn't really a fair representation of Dweck's own views: she is not claiming that a growth mind-set can work miracles where there is no aptitude, simply that it is one of many important elements, particularly when we find ourselves facing new challenges that would cause us to question our talents. Common sense would suggest that there is still a threshold of intelligence that is necessary for success, but your mind-set makes the difference in whether you can capitalize on that potential when you are outside of your comfort zone.

Some people also cite the growth mind-set as a reason to rhapsodise over a child's every achievement and ignore their flaws. In reality, her message is quite the opposite: overpraising a child for effort or success may be almost as damaging as scolding them for failure. Telling a child that "you're smart" after a good result, for example, appears to reinforce a fixed mind-set. The child may begin to feel embarrassed if they put a lot of effort into their studies—since that would detract from their smartness. Or they may avoid future challenges that might threaten to take them down off this pedestal. Ironically, Eddie Brummelman at the University of Amsterdam has found that excessive praise can be particularly damaging to children with low self-esteem, who may become scared of failing to live up to parental expectations in the future.[31]

We shouldn't avoid showing pride in a child's achievements, of course; nor should we shy away from offering criticism when they have failed. In each case, the researchers advise that parents and teachers emphasise the journey that led to their goal, rather than the result itself.[32] As Dweck explains, "It is about telling the truth about a student's current achievement and then, together, doing something about it, helping him or her become smarter."

Sara Blakely, the founder of the intimate clothes company Spanx, offers us one example of this principle in action. Describing her childhood, she recalled that every evening after school her father would ask her, "What did you fail at today?" Out of context, it might sound cruel, but Blakely understood what he meant: if she hadn't failed at anything, it meant that she hadn't stepped out of her comfort zone, and she was limiting her potential as a result.

"The gift he was giving me is that failure is [when you are] not trying versus the outcome. It's really allowed me to be much freer in trying things and spreading my wings in life," she told CNBC. That growth mind-set, combined with enormous creativity, eventually allowed her to ditch her job selling fax machines to invest $5,000 in her own business. That business is now worth more than a billion dollars.[33]

Dweck has recently been exploring relatively brief mind-set interventions that could be rolled out on a large scale, finding that an online course teaching schoolchildren about neuroplasticity—the brain's ability to rewire itself—reduces the belief that intelligence and talent are fixed, innate qualities.[34] In general, however, the average long-term benefits of these one-shot interventions are significant but modest,[35] and more profound change would almost certainly require regular reminders and deliberate consideration from everyone involved.

The goal, ultimately, is to appreciate the process rather than the end result—to take pleasure in the act of learning even when it's difficult. And that itself will take work and perseverance, if you have spent your whole life believing that talent is purely innate and success should come quickly and easily.

~

In light of all these findings, Feynman's astonishing personal development—from tinkering schoolboy to world-class scientist—begins to make a lot of sense.

From early childhood, he was clearly overflowing with an irrepressible desire to understand the world around him—a trait that he learned from his father. "Wherever we went there were always new wonders to hear about; the mountains, the forests, the sea."[36]

With this abundant curiosity, he needed no other motivation to study. As a student, it would drive him to work all night on a problem—for the sheer pleasure of finding an answer—and as a working scientist, it allowed him to overcome professional frustrations.

When first arriving as a professor at Cornell, for instance, he began to fear that he could never live up to his colleagues' expectations; he began to suffer burnout, and the very thought of physics began to "disgust" him. Then he remembered how he had once "played" with physics as if it were a toy. He determined from that point on to experiment with only the questions that actually interested him—no matter what others might think.

Just when many would have lost their curiosity, he had reignited it once more—and that continued desire to "play" with complex ideas would ultimately lead to his greatest discovery. In the cafeteria of Cornell, he watched a man throwing plates in the air and catching them. Feynman was puzzled by their movement—the way they wobbled, and how that related to the speed at which they were spinning. As he put that motion into equations he began to see some surprising parallels with an electron's orbit, eventually leading to his influential theory of quantum electrodynamics that won a Nobel Prize. "It was like uncorking a bottle," he later said. "The diagrams and the whole business that I got the Nobel Prize for came from that piddling around with the wobbling plate."[37]

"Imagination reaches out repeatedly trying to achieve some higher level of understanding, until suddenly I find myself momentarily alone before one new corner of nature's pattern of beauty and true majesty revealed," he added. "That was my reward."[38]

Along the way, he was aided by a growth mind-set that allowed him to cope with failure and disappointment—beliefs he passionately expressed in his Nobel lecture. "We have a habit of writing articles published in scientific journals to make the work as finished as possible, to cover all the tracks, to not worry about the blind alleys or to describe how you had the wrong idea first, and so on," he said. Instead, he wanted to use the lecture to explain the challenges he had faced, including "some of the unsuccessful things on which I spent almost as much effort, as on the things that did work."

He describes how he had been blind to apparently fatal flaws in his initial theory, which would have resulted in physical and mathematical impossibilities, and he was remarkably candid about his disappointment when his mentor pointed out these defects. "I suddenly realized what a stupid fellow I am." Nor did the resolution of these difficulties come from single flash of genius; the moments of inspiration were separated by long periods of "struggle." (He repeated the word six times during the speech.)

His colleague Mark Kac may have considered him "a magician of the highest calibre," an "incomprehensible" genius, but he took an earthier view of himself. Unlike many other high achievers, he was willing to acknowledge the blood, sweat, and tears, and sometimes tedious drudgery, that he had faced for the sheer "excitement of feeling that possibly nobody has yet thought of the crazy possibility you are looking at right now."[39]

~

By enhancing our learning and pushing us to overcome failures in these ways, curiosity and the growth mind-set would already constitute two important mental characteristics, independent of general intelligence, that can change the path of our lives. If you want to make the most of your intellectual potential, they are essential qualities that you should try to cultivate.

But their value does not end here. In an astonishing convergence with the theories of evidence-based wisdom, the very latest research shows that both curiosity and the growth mind-set can also protect us from the dangerously dogmatic, one-sided reasoning that we explored in earlier chapters. The same qualities that will make you *learn* more productively also make you *reason* more wisely, and vice versa.

To understand why, we first need to return to the work of Dan Kahan at Yale University. As you may recall, he found intelligence and education can exaggerate "motivated reasoning" on subjects such as climate change—leading to increasingly polarized views.

Those experiments had not considered the participants' natural interest, however, and Kahan was curious to discover whether a hunger for new information might influence the ability to assimilate alternative viewpoints.

To find out, he first designed a scale that tested his participants' science curiosity, which included questions about their normal reading habits (whether they would read about science for pleasure), whether they kept up to date with scientific news, and how often they would talk about science with friends or family. Strikingly, he found that some people had a large knowledge but low curiosity—and vice versa. And that finding would be crucial for explaining the next stage of the experiment, when Kahan asked the participants to give their views on politically charged subjects such as climate change.

As he had previously shown, greater knowledge of science only increased polarization between left and right. But this was not true for curiosity, which reduced the differences. Despite the prevailing views of most conservative thinkers, more curious Republicans were more likely to endorse the scientific consensus on global warming, for instance.

It seemed that their natural hunger for understanding had overcome their prejudices, so that they were readier to seek out material that challenged their views. Sure enough, when given the choice between two articles, the more curious participants were more willing to read a piece that challenged rather than reinforced their ideology. "They displayed a marked preference for novel information, even when it was contrary to their political predispositions," Kahan wrote in the accompanying paper.[40] In other words, their curiosity allowed the evidence to seep through those "logic-tight compartments" that normally protect the beliefs that are closest to our identities.

Kahan admits to being "baffled" by the results; he told me that he had fully expected that the "gravitational pull" of our identities would have overpowered the lure of curiosity. But it makes sense when you consider that curiosity helps us to tolerate uncertainty. Whereas incurious people feel threatened by surprise, curious people relish the mystery. They enjoy being taken aback; finding out something new gives them that dopamine kick. And if that new information raises even more questions, they'll rise to the bait. This makes them more open-minded and willing to change their opinions, and stops them becoming entrenched in dogmatic views.

In ongoing research, Kahan has found similar patterns for opinions on issues such as firearm possession, illegal immigration, the legalization of marijuana, and the influence of pornography. In each case, the itch to find out something new and surprising helped to counteract the politically motivated reasoning.[41]

Further cutting-edge studies reveal that the growth mind-set can protect us from dogmatic reasoning in a similar way, by increasing our intellectual humility. Studying for a doctorate under the supervision of Carol Dweck at Stanford University, Tenelle Porter first designed and tested a scale of intellectual humility, asking participants to rate statements such as "I am willing to admit if I don't know something," "I actively seek feedback on my ideas, even if it is critical" or "I like to compliment others on their intellectual strengths." To test whether their answers reflect their behavior, Porter showed that their scores also corresponded to the way they react to disagreement on issues like gun control—whether they would seek and process contradictory evidence.

She then separated the participants into two groups. Half read a popular science article that emphasised the fact that our brains are malleable and capable of change, priming the growth mind-set, while the others read a piece that described how our potential is innate and fixed. Porter then measured their intellectual humility. The experiment worked exactly as she had hoped: learning about the brain's flexibility helped to promote a growth mind-set, and this in turn produced greater humility, compared to those who had been primed with the fixed mind-set.[42]

Porter explained it to me like this: "If you have the fixed mind-set, you are all the time trying to find out where you stand in the hierarchy; everyone's ranked. If you're at the top, you don't want to fall or be taken down from the top, so any sign or suggestion that you don't know something or that someone knows more than you—it's threatening to dethrone you." And so, to protect your position, you become overly defensive. "You dismiss people's ideas with the notion that 'I know better so I don't have to listen to what you have to say.'"

In the growth mind-set, by contrast, you're not so worried about proving your position relative to those around you, and your knowledge doesn't represent your personal value. "What's more, you are motivated to learn because it makes you smarter, so it is a lot easier to admit what you don't know. It doesn't threaten to pull you down from any kind of hierarchy."

Igor Grossmann, incidentally, has come to similar conclusions in one of his most recent studies, showing that the growth mind-set is positively correlated with his participants' everyday wise reasoning scores.[43]

Feynman, with his curiosity and growth mind-set, certainly saw no shame in admitting his own limitations—and welcomed this intellectual humility in others. "I can live with doubt, and uncertainty, and not knowing. I think it's much more interesting to live not knowing anything than to have answers which might be wrong," he told the BBC in 1981. "I have approximate answers, and possible beliefs, and different degrees of certainty about different things, but I'm not absolutely sure of anything."[44]

This was also true of Benjamin Franklin. He was famously

devoted to the development of virtues, seeing the human mind as a malleable object that could be molded and honed. And his many "scientific amusements" spanned the invention of the electric battery, the contagion of the common cold, the physics of evaporation and the physiological changes that come with exercise. As the historian Edward Morgan put it: "Franklin never stopped considering things he could not explain. He could not drink a cup of tea without wondering why tea leaves gathered in one configuration rather than another."[45] For Franklin, like Feynman, the reward was always in the discovery of new knowledge itself, and without that endlessly inquisitive attitude, he may have been less open-minded in his politics too.

And Darwin? His hunger to understand did not end with the publication of *On the Origin of Species,* and he maintained a lengthy correspondence with skeptics and critics. He was capable of thinking independently while also always engaging with and occasionally learning from others' arguments.

These qualities may be more crucial than ever in today's fast-moving world. As the journalist Tad Friend noted in the *New Yorker*: "In the nineteen-twenties, an engineer's 'half-life of knowledge'—the time it took for half of his expertise to become obsolete—was thirty-five years. In the nineteen-sixties, it was a decade. Now it's five years at most, and, for a software engineer, less than three."[46]

Porter agrees that children today need to be better equipped to update their knowledge. "To learn well may be more important than knowing any particular subject or having any particular skill set. People are moving in and out of different careers a lot, and because we're globalizing we are exposed to lots of different perspectives and ways of doing things."

She points out that some companies, such as Google, have already announced that they are explicitly looking for people who combine passion with qualities like intellectual humility, instead of traditional measures of academic success like a high IQ or Grade Point Average. "Without humility, you are unable to learn," Laszlo Bock, the senior vice president of people operations for Google, told the *New York Times*.[47]

"Successful bright people rarely experience failure, and so they don't learn how to learn from that failure," he added. "They, instead, commit the fundamental attribution error, which is if something good happens, it's because I'm a genius. If something bad happens, it's because someone's an idiot or I didn't get the resources or the market moved. . . . What we've seen is that the people who are the most successful here, who we want to hire, will have a fierce position. They'll argue like hell. They'll be zealots about their point of view. But then you say, 'here's a new fact," and they'll go, 'Oh, well, that changes things; you're right.'"

Bock's comments show us that there is now a movement away from considering SAT scores and the like as the sum total of our intellectual potential. But the old and new ways of appraising the mind do not need to be in opposition, and in Chapter 8 we will explore how some of the world's best schools already cultivate these qualities and the lessons they can teach us all about the art of deep learning.

~

If you have been inspired by this research, one of the simplest ways to boost anyone's curiosity is to become more autonomous during learning. This can be as simple as writing out what you already know about the material to be studied and then setting down the questions you really want to answer. The idea is to highlight the gaps in your knowledge, which is known to boost curiosity by creating a mystery that needs to be solved, and it makes it personally relevant, which also increases interest.

It doesn't matter if these are the same questions that would come up in an exam, say. Thanks to the spillover effect from the dopamine kick, you are more likely to remember the other details too, with some studies revealing that this small attempt to spark your engagement can boost your overall recall while also making the whole process more enjoyable. You will find that you have learned far more effectively than if you had simply studied the material that you believe will be most *useful*, rather than interesting.

The wonderful thing about this research is that learning seems to beget learning: the more you learn, the more curious you become, and the easier it becomes to learn, creating a virtuous cycle. For this reason, some researchers have shown that the best predictor of how much new material you will learn—better than your IQ—is how much you already know about a subject. From a small seed, your knowledge can quickly snowball. As Feynman once said, "everything is interesting if you go into it deeply enough."

If you fear that you are too old to reignite your curiosity, you may be interested to hear about Feynman's last great project. As with his famous Nobel Prize–winning discoveries, the spark of interest came from a seemingly trivial incident. During a dinner in the summer of 1977, Feynman's friend Ralph Leighton had happened to mention a geography game in which each player had to name a new, independent country.

"So you think you know every country in the world?" Richard cheekily responded. "Then whatever happened to Tannu Tuva?" He remembered collecting a stamp from the country as a child; it was, he said, a "purple splotch on the map near Outer Mongolia." A quick check in the family atlas confirmed his memory.

It could have ended there, but the lure of this unknown country soon became something of an obsession for the two men. They listened to Radio Moscow for any mentions of this obscure Soviet region, and scoured university libraries for records of anthropological expeditions to the place, which offered them small glimpses of the country's beautiful saltwater and freshwater lakes in its countryside, its haunting throat singing and shamanic religion. They discovered that the capital, Kyzyl, housed a monument marking it as the "Center of Asia"—though it was unclear who had built the statue—and that the country was the source of the Soviet Union's largest uranium deposit.

Eventually Leighton and Feynman found a Russian-Mongolian-Tuvan phrasebook, which a friend helped to translate into English, and they began writing letters to the Tuvan Scientific Research Institute of Language, Literature, and History, asking for a cultural exchange. However, each time they thought they had a chance of reaching their goal, they were rebuffed by Soviet bureaucracy—but they persevered anyway.

By the late 1980s, Feynman and Leighton believed they had finally found a way into the country: on a trip to Moscow, Leighton managed to arrange for a Soviet exhibition of Eurasian nomadic cultures to visit the US, and as part of his role as an organizer, he bargained a research and filming trip to Tuva. The exhibition opened at the Natural History Museum of Los Angeles in February 1989, and it was a great success, introducing many more people to a culture that remains little known in the West.

Feynman, alas, never lived to see the country; he died on February 15, 1988, from abdominal cancer, before his longed-for trip could be arranged. Right until the end, however, his passion continued to animate him. "When he began talking about Tuva, his malaise disappeared," Leighton noted in his memoir, *Tuva or Bust*. "His face lit up, his eyes sparkled, his enthusiasm for life was infectious." Leighton recalls walking the streets with Feynman after one round of surgery, as they tested each other on Tuvan phrases and imagined themselves taking a turn through Kyzyl—a way of building Feynman's strength and distracting him from his discomfort.

And in his last years, Feynman had piqued the interest of many others, resulting in a small organization, the Friends of Tuva, being founded to share his fascination; Feynman's curiosity had built a small bridge across the Iron Curtain. When Leighton finally reached Kyzyl himself, he left a small plaque in memory of Feynman, and his daughter Michelle would make her own visit in the late 2000s. "Like [Ferdinand] Magellan, Richard Feynman completed his last journey in our minds and hearts," Leighton wrote in his memoir. "Through his inspiration to others, his dream took on a life of its own."

8

The benefits of eating bitter: *East Asian education and the three principles of deep learning*

James Stigler's heart was racing and his palms were sweaty—and he wasn't even the one undergoing the ordeal.

A graduate student at the University of Michigan, Stigler was on his first research trip to Japan, and he was now observing a fourth-grade lesson in Sendai. The class were learning how to draw three-dimensional cubes, a task that is not as easy as it might sound for many children, and as the teacher surveyed the students' work, she quickly singled out a boy whose drawings were particularly sloppy and ordered him to transfer his efforts to the blackboard—in front of everyone.

As a former teacher himself, Stigler found that his interest immediately piqued. Why would you choose the worst student—rather than the best—to demonstrate their skills, he wondered? It seemed like a public humiliation rather than a useful exercise.

The ordeal didn't end there. After each new attempt the boy made, his teacher would ask the rest of the class to judge whether his drawings were correct—and when they shook their heads, he had to try again. The boy ended up spending forty-five minutes standing at that blackboard, as his failings continued to be exposed to everyone around him.

Stigler could feel himself becoming more and more uncomfortable for the poor child. "I thought it was like torture." He was sure that the boy was going to break down in tears any moment. In the US, Stigler knew that a teacher could even be fired for treating a child this way. The boy was only around nine or ten, after all. Wasn't it cruel to explore his flaws so publicly?[1]

~

If you were brought up in a Western country, you will have probably shared Stigler's reactions as you read this story: in European and American cultures, it seems almost unthinkable to discuss a child's errors so publicly, and only the very worst teachers would dream of doing such a thing. You may even think that it only underlines some of the serious flaws in East Asian education systems.

It is well known, true, that countries such as Japan, Mainland China, Taiwan, Hong Kong, and South Korea regularly outperform many Western countries on measures of education, such as PISA (the Program for International Student Assessment). But there is also a widely held suspicion among Western commentators that those achievements are largely the result of severe classroom environments. East Asian schools, they argue, encourage rote learning and discipline, at the expense of creativity, independent thinking, and the child's own well-being.[2]

Stigler's observation in Sendai would, at first, only seem to confirm those suspicions. As he continued his studies, however, he found that these assumptions about East Asian education are completely unfounded. Far from relying on dry, rote memorization, it turns out that Japanese teaching strategies automatically encourage many of the principles of good reasoning, such as intellectual humility and actively open-minded thinking, that can protect us from bias, while also improving factual learning. In many ways it is the Western system of education—particularly in the US and UK—that stifles flexible, independent thinking, while also failing to teach the factual basics.

In this light, it becomes clear that the teacher's behavior in Sendai—and the boy's struggles at the blackboard—happen to reflect the very latest neuroscience on memory.

Building on the research we explored in the last chapter, these cross-cultural comparisons reveal some simple practical techniques that will improve our mastery of any new discipline while also offering further ways that schools could help young thinkers to avoid the intelligence trap.

Before we return to that classroom in Sendai, you may want to consider your own intuitions about the way you personally learn, and the scientific evidence for or against those beliefs.

Imagine you are learning any new skill—piano, a language, or a professional task—and then decide whether you agree or disagree with each of the following statements:

- The more I improve my performance today, the more I will have learned.
- The easier material is to understand, the more I will memorize.
- Confusion is the enemy of good learning and should be avoided.
- Forgetting is always counterproductive.
- To improve quickly, we should only learn one thing at a time.
- I remember more when I feel like I am struggling than when things come easily.

Only the last statement is supported by neuroscience and psychology, and the rest all reflect common myths about learning.

Although these beliefs are related to Carol Dweck's work on the growth and fixed mind-sets, they are also quite different. Remember that a growth mind-set concerns the beliefs about yourself, and whether your talents can improve over time. And while that may mean that you are more likely to embrace challenge *when necessary*, it's perfectly possible—and indeed likely—that you could have a growth mind-set without necessarily thinking that confusion and frustration will, in and of themselves, improve your learning.

The latest neuroscience, however, shows that we learn best when we are confused; deliberately limiting your performance today actually means you will perform better tomorrow. And a failure to recognize this fact is another primary reason that many people—including those with high IQs—often fail to learn well.

One of the earliest studies to show this phenomenon was commissioned, strangely enough, by the British Post Office, which in the late 1970s asked the psychologist Alan Baddeley to determine the best methods to train its staff.*

The Post Office had just invested heavily in machines that could sort letters according to postcodes, but to use these machines, their 10,000 postal workers had to learn how to type and use a keyboard, and Baddeley's job was to discover the most efficient schedule for training.

The assumption of many psychologists at the time had been that intensive training would be better—the postmen should be allowed to devote a few hours a day to mastering the skill. And this was indeed the way that the workers themselves preferred: they were able to see real progress during that time; at the end of their sessions, their typing felt much more fluent than at the beginning, and they assumed this carried over into long-term memory.

For comparison, however, Baddeley also created a few groups that learned in shorter stretches, over a longer period of time: just one hour a day, compared to four. The workers in this group didn't seem to like the approach; they lacked the sense of mastery at the end of their session and didn't feel like they were progressing as quickly as those taking the longer sessions.

But they were wrong. Although the sessions themselves felt unsatisfying compared to those of the people who developed more proficiency in a single day, these subjects ended up learning and remembering much more in relation to the amount of time they were putting into it. On average, a person with the "spaced" approach mastered the basics within thirty-five hours, compared to fifty hours for the intensive learners—a 30 percent difference. Individually, even the *slowest* person in the one-hour-a-day group had mastered the skill in less time than the *quickest* learner among those devoting four hours a day to the test. When the researchers followed up their stud-

---

* Baddeley, incidentally, is also the reason that British postcodes contain around six or seven characters—which is the maximum span of human working memory. He also advised on the specific positioning of the numbers and the letters to make them as memorable as possible.

ies a few months later, the spaced learners were still quicker and more accurate than those learning in blocks.[3]

Today, the spacing effect is well known to psychological scientists and many teachers, and it is often represented as demonstrating the benefits of rest and the dangers of cramming. But the true mechanism is more counter-intuitive, and hinges on the very frustration that had annoyed the postmen.

By splitting our studies into smaller chunks, we create periods in which we can forget what we've learned, meaning that at the start of the next session, we need to work harder to remember what to do. That process—of forgetting, and then forcing ourselves to relearn the material—strengthens the memory trace, leading us to remember more in the long term. People who learn in longer blocks miss out on those crucial steps—the intermediate forgetting and relearning—that would promote long-term recall precisely because it is harder.

In this way, Baddeley's study gave some of the first hints that our memory can be aided by "desirable difficulties"—additional learning challenges that initially impair performance, but which actually promote long-term gains.

The neuroscientists Robert and Elizabeth Bjork at the University of California, Los Angeles, have pioneered much of this work, showing that desirable difficulties can be powerful in many different circumstances—from math to foreign languages, art history, musical performance, and sports.

Consider a physics class revising for an exam. In the West, it's common for teachers to present the principles and then to get students to repeat endless series of similar questions until they reach nearly 100 percent accuracy. The Bjorks have shown that, in reality, learning is more effective if the student solves just enough problems to refresh their mind, before moving on to a new (perhaps related) subject, and only later should they return to the initial topic.

Like the spacing effect, this process of switching between tasks—known as interleaving—can lead the student to feel confused and overburdened, compared to lessons in which they are allowed to focus solely on one subject. But when they are tested later, they have learned much more.[4]

Other desirable difficulties include "pre-testing" or "productive failure"—in which students are quizzed on facts they haven't yet learned, or given complex problems they don't yet know how to solve.

You can see it for yourself. Without looking at the answers below, try to match the following Italian words to the English terms.

- I pantaloni
- L'orologio
- La farfalla
- La cravatta
- Lo stivale

- Tie
- Trousers
- Boot
- Bowtie
- Watch

Now look at the footnote below to see the answers.*

You might be able to guess a couple, perhaps from their similarity to English or French terms, or from previous exposure to the language. But the surprising thing about pre-testing is that it doesn't matter if your initial answers are right or if you really don't have a clue; it's the act of thinking that will boost your learning. Like the fruitful forgetting that comes with spaced study, the frustration we feel at not understanding these terms leads us to encode the information more deeply, and even if you are not normally very good at learning languages, you should find that this vocabulary ultimately sticks in your mind.[5]

(This was, incidentally, the reason that I asked you to guess the truth of those statements at the start of this chapter. By causing you to question your existing knowledge, it should enable you to remember the subsequent information far more clearly.)

Productive failure seems to be particularly fruitful for disciplines like math, in which teachers may ask students to solve problems *before* they've been explicitly taught the correct methods. Studies

---

* I pantaloni—trousers
L'orologio—watch
La farfalla—bowtie
La cravatta—tie
Lo stivale—boot

suggest they'll learn more and understand the underlying concepts better in the long run, and they will also be better able to translate their learning to new, unfamiliar problems.[6]

The introduction of desirable difficulties could also improve our reading materials. Textbooks that condense concepts and present them in the most coherent and fluent way possible, with slick diagrams and bullet point lists, actually *reduce* long-term recall. Many students—particularly those who are more able—learn better if the writing is more idiosyncratic and nuanced, with a greater discussion of the potential complications and contradictions within the evidence. People reading the complex prose of Oliver Sacks remembered more about the visual perception than people looking at a slick, bullet-pointed textbook, for instance.[7]

In each case, we can see how the elements of confusion that our whole education system is geared to avoid would lead to more profound thinking and learning, if only we let students feel a bit frustrated.

"Current performance is a measure of accessibility [of information] right now, but learning is about more fundamental changes that will be reflected after a delay, or a transfer of this learning to somewhere else," Robert told me when I met him and his wife in the faculty center of UCLA. "So if you interpret current performance as a measure of learning, you will get lots of things wrong."

Scientifically speaking, these results are no longer controversial. The evidence is now unarguable: introducing desirable difficulties into the classroom—through strategies such as spacing, interleaving, and productive failure—would ensure that everyone learns more effectively.

Unfortunately, it is hard to persuade people to appreciate these effects; like Baddeley's postal workers, students, parents, and even many teachers still assume that the easier you find it to learn something today, the better you will perform tomorrow, even though these assumptions are deeply flawed. "We have all these results showing that people prefer the poorer ways of learning," Robert added.[8] "So you're not going to make your students happier right away."

Elizabeth agreed. "They interpret [confusion] negatively, as opposed to thinking that this is an opportunity to learn something or understand it in a better way."

It's as if we went to the gym hoping to build our muscles, but then only ever decided to lift the lightest weights. The Bjorks have found that these "metacognitive illusions" are surprisingly resilient even after people have seen the evidence or experienced the benefits for themselves. The result is that disappointingly few schools try to make use of desirable difficulties, and millions of students are suffering as a result, when they could be learning much more effectively, if only they knew how to embrace confusion.

~

At least, that's the view we get from the US and the UK.

But as we have seen previously, we should be wary of assuming that the biases documented among "WEIRD" countries represent human universals, when studies of East Asian cultures often show very different attitudes.

Rather than us feeling that learning should come easily, various surveys—including Stigler's own research—show that students in countries such as Japan appreciate that struggle is necessary in education. If anything, the students in these cultures are concerned if the work isn't hard enough.

These notions can be found in the parents' and teachers' attitudes, in popular sayings such as "*doryoki ni masaru, tensai nashi*" ("even genius cannot transcend effort") and in the country's folk stories. Most Japanese schoolchildren will have heard the story of nineteenth-century scholar Ninomiya Sontoku, for instance. As a poor young boy, he was said to use every opportunity for study, even when collecting firewood in the forest, and many schoolyards still contain statues of Sontoku with a book in front of his nose and firewood on his back. From earliest infancy, children in Japan are immersed in a culture that tolerates struggle and challenge.

Crucially, this acceptance of struggle extends to their mind-sets about their own talents: Japanese students are more likely to see their abilities as a work in progress, leading to the growth mind-set. "It's not that the Japanese don't believe in individual differences," Stigler told me. "It's just that they don't view them as limitations to the extent that we do." As a result, an error or mistake is not seen as a sign of some kind of permanent, inevitable failure—it is, according to Stigler, "an index of what still needs to be learned."

These beliefs help to explain why East Asian students, in general, are willing to work longer hours, and even if they aren't naturally gifted, they are more likely to try to make up for that by studying harder. Just as importantly, however, these beliefs also have consequences for the ways the teachers handle their lessons, allowing them to introduce more desirable difficulties into the curriculum. The result is that their pedagogical methods regularly embrace confusion to enhance learning and comprehension.

When tackling a new topic in math or science, for instance, it's quite common for Japanese teachers to begin their lessons by asking students to solve a problem *before* they've been told the exact method to apply—the use of "productive failure" that we discussed a few pages ago. The next couple of hours are then devoted to working their way through those challenges—and although the teacher offers some guidance, the bulk of the work is expected to come from the students.

American or British students balk at the confusion this produces—and when they begin to struggle, the teacher would be tempted to give in and tell them the answer.[9] But Stigler found that Japanese students relished the challenge. As a consequence, they think more deeply about the underlying characteristics of the problem, increasing their ultimate understanding and their long-term recall.

And when working through problems, they are also encouraged to consider alternative solutions, besides the most obvious one, and fully to explore their mistakes (together with those of their classmates) to understand how one approach works and another doesn't. From a very young age, they are encouraged to take a more holistic view of problems, and to see the underlying connections between different ideas. As one elementary maths teacher in Stigler's studies put it: "We face many problems every day in the real world. We have to remember that there is not only one way we can solve each problem."

British or American schools, in contrast, often discourage that exploration, for fear that it might cause extra confusion; for each type of math or science problem, say, we are only taught one potential strategy to find a solution. But the scientific research shows that comparing and contrasting different approaches gives a better understanding of the underlying principles—even if it does result in more confusion at the beginning.

"They've created a classroom culture that tries to support the prolonged extension of confusion," Stigler said. "The Japanese feel that if you can sustain that in the classroom, students will learn more. Whereas we're very focused on [simply] getting the answer—and if you want the students to get the right answer, you then make it as easy as possible."

In this light, our reactions to Stigler's story of the Japanese boy struggling to draw 3D cubes at the blackboard make more sense. Whereas Americans, or Brits, would see the boy's initial troubles as a sign of weakness or stupidity, his Japanese classmates saw his perseverance. "His errors were not a matter of great concern; what would be worrisome would be the child's failure to expend the effort necessary to correct them." The boy didn't cry, as Stigler had initially expected, because in that cultural context there simply wasn't any reason to feel the same level of personal shame that we would expect.

On the contrary, Stigler says that as the lesson went on, he felt a wave of "kindness from his classmates and the teacher." "No one was going to cut him a break to say that was good enough until he'd accomplished the job. But on the other hand, you had the sense that they were all there to help him," Stigler told me. They all knew that the struggle was the only way the boy was going to learn and catch up with his peers.

It is true that some lessons do involve elements of rote learning, to ensure that the basic facts are memorized and easily recalled, but this research shows that Japanese classrooms provide far more room for independent thinking than many Western commentators have assumed. And the benefits can be seen not only in the PISA scores, but also in tests of creative problem solving and flexible thinking, in which Japanese students also outperform the UK and US, demonstrating that these students are also better equipped to transfer their knowledge to new and unexpected tasks.[10]

Although some of the strongest evidence comes from Japan, the value of struggle does seem to be common to other Asian teaching cultures too, including Mainland China, Hong Kong, and Taiwan. In Mandarin, for instance, the concept of *chiku* or "eating bitterness" describes the toil and hardship that leads to success. And Stigler has since expanded his focus to explore other countries, such as the Netherlands, which has also outperformed the US and UK. Although they may vary on many factors, such as the class size, or the specific methods that the teachers use to deliver material, the top-performing schools all encourage students to go through those periods of confusion.

Stigler has now been researching these ideas for decades, and he suggests his findings can be distilled to three stages of good teaching:[11]

- **Productive struggle:** Long periods of confusion as students wrestle with complex concepts beyond their current understanding.

- **Making connections:** When undergoing that intellectual struggle, students are encouraged to use comparisons and analogies, helping them to see underlying patterns between different concepts. This ensures that the confusion leads to a useful lesson—rather than simply ending in frustration.

- **Deliberate practice:** Once the initial concepts have been taught, teachers should ensure that students practice those skills in the most productive way possible. Crucially, this doesn't involve simply repeating near identical problems ad nauseam, as you might find in the Western maths classroom, but means adding additional variety and challenges—and yet more productive struggle.

These are profound discoveries that offer some of the most robust ways of improving education and academic achievement, and in a few pages we'll see how anyone can make use of desirable difficulties to master new skills.

But these findings are not only interesting for what they tell us about human memory; I believe that they also reveal some profound insights into the cultural origins of the intelligence trap.

If you consider classrooms in the UK and US, for instance, our mental worth is often judged by who can put their hand up quickest—giving us the subtle signal that it's better to go with an immediate intuitive response without reflecting on the finer details. And you are not going to be rewarded for admitting that you don't know the answer; intellectual humility is actively discouraged.

Worse still, the lessons are often simplified so that we can digest the material as quickly as possible—leading us to prefer "fluent" information over material that might require deeper consideration. Particularly in earlier education, this also involves glossing over potential nuances, such as the alternative interpretations of evidence in history or the evolution of ideas in science, for instance—with facts presented as absolute certainties to be learned and memorized.[12] The assumption had been that introducing these complexities would be too confusing for younger students—and although the teaching methods do allow more flexibility at high school and university, many students have already absorbed a more rigid style of thinking.

Even some well-meaning attempts at educational reform fall into these traps. Teachers have been encouraged to identify a child's learning style—whether they are a visual, verbal, or kinesthetic learner. The idea sounds progressive, but it only reinforces the idea that people have fixed preferences for the ways they learn, and that we should make learning as easy as possible, rather than encouraging them to wrestle with problems that aren't immediately straightforward.

It's little wonder that students in countries such as the US and UK do not tend to score well on Igor Grossmann's tests of evidence-based wisdom, or the measures of critical thinking that predict our susceptibility to misinformation.

Now compare those attitudes to the Japanese education system, where even students in elementary school are encouraged to wrestle with complexity every day; they are taught to discover new ways of solving problems for themselves and, when they have found one answer, to consider the other alternative solutions. If you don't immediately understand something, the answer is not to ignore it and reinforce your own beliefs, but to look further and to explore its nuances. And the extra thinking that involves is not a sign of weakness or stupidity; it means that you are capable of "eating bitterness" to come to a deeper understanding. If you initially fail, it's fine to admit your mistakes, because you know you can improve later.

The students are simply better prepared for the more complex, nuanced, and ill-defined problems the real world will set against them during adulthood. And this seems to be reflected in their higher scores on measures of open-minded, flexible reasoning.[13] Various studies, for instance, have found that when asked about controversial environmental or political issues, people in Japan (and other East Asian cultures) tend to take longer to consider the questions without offering knee-jerk reactions, and are more likely to explore contradictory attitudes and to think about the long-term consequences of any policies.[14]

If we return to that idea of the mind as a car, the British and American education systems are designed to offer as smooth a track as possible, so that each person can drive as fast as their engine can possibly let them. The Japanese education system, in contrast, is more of an assault course than a race course; it requires you to consider alternative routes to steer your way around obstacles and persevere even when you face rough terrain. It trains you to navigate effectively rather than simply revving the engine.

Let's be clear: we are talking about averages here, and there is a huge amount of variation within any culture. But these results all suggest that the intelligence trap is partly a cultural phenomenon born in our schools. And once you recognize these facts, it becomes clear that even small interventions can begin to encourage the thinking styles we have explored in the rest of this book, while also improving the factual, academic learning that schools already try to cultivate.

Even a simple strategic pause can be a powerful thing.

Having asked the class a question, the average American teacher typically waits less than a second before picking a child to provide an answer—sending out a strong message that speed is valued over complex thinking. But a study from the University of Florida has found that something magical happens when the teacher takes a little more time—just three seconds—to wait to pick a child, and then for the child to think about the response.

The most immediate benefit was seen in the length of the children's answers. The small amount of thinking time meant that the children spent between three and seven times as long elaborating their thoughts, including more evidence for their viewpoint and a greater consideration of alternative theories. The increased waiting time also encouraged the children to listen to each other's opinions and develop their ideas. Encouragingly, their more sophisticated thinking also translated to their writing, which became more nuanced and complex. That's an astonishing improvement from the simple act of exercising teacherly patience.[15] As the researcher, Mary Budd Rowe, put it in her original paper: "slowing down may be a way of speeding up."

The psychologist Ellen Langer at Harvard University, meanwhile, has examined the way that complex material is currently oversimplified to avoid any ambiguity, and the consequences of that for our thinking. In physics or mathematics, for instance, there may be many different ways to solve a problem—but we are told just one method and discouraged from looking beyond that. The assumption had been that even a hint of complexity would only lead to confusion—which was thought to harm learning. Why let the child potentially mix up the different methods if just one would do?

In reality, Langer has found that subtly changing the phrasing of a lesson to introduce those ambiguities encourages deeper learning. In one high-school physics lesson, the children were presented with a thirty-minute video demonstrating some basic principles and asked to answer some questions using the information. In addition to the basic instructions, some of the participants were told that "The video presents only one of several outlooks on physics, which may or may not be helpful to you. Please feel free to use any additional methods you want to assist you in solving problems." This simple prompt encouraged them to think more freely about the topic at hand, and to apply the material more creatively to novel questions.[16]

In another experiment, students were offered a strategy to solve a specific kind of mathematical problem. Thanks to a change in a single word, children told that this was just "*one* way to solve this equation" performed better than those told it was "*the* way to solve this equation"—they were about 50 percent more likely to get the correct answer. They also showed a deeper understanding of the underlying concept and were better able to determine when the strategy would and wouldn't work.[17] The same applies for the humanities and social sciences. Geography students told that "this *may* be the cause of the evolution of city neighborhoods" showed a greater understanding in a subsequent test than those who had been taught the material as absolute, undeniable facts.

The subtle suggestion of ambiguity, far from creating confusion, invites them to consider the other alternative explanations and explore new avenues that would have otherwise been neglected. The result is a more reflective and actively open-minded style of thinking, of the kind that we explored in Chapter 4. Framing questions using conditional terms can also improve students' performance on creative thinking tasks.

You may remember that Benjamin Franklin deliberately avoided the use of "dogmatic" terms that demonstrated complete certainty, and that this acceptance of uncertainty also improved the superforecasters' decision making. Langer's work offers further evidence that this kind of nuanced thinking can be encouraged from a young age.[18] Presenting uncertainty might create a little bit of confusion—but that only improves engagement and ultimately increases the child's learning.

In addition to these subtle primes, students might be actively encouraged to try to imagine a historical article from various viewpoints and the arguments they might raise, for instance. Or in science, they might be given two case studies representing apparently contradictory arguments on the topic they are studying, and then asked to evaluate the strength of the evidence and reconcile the different views. Again, the prediction might have been that these exercises would be counterproductive, a distraction that reduces students' overall learning of the syllabus. In reality, they add another desirable difficulty that means children actually remember more of the factual material than those who are specifically told to memorize the text.[19]

If you combine these methods with the measures we have explored previously, such as the training in emotion differentiation that we examined in Chapter 5, and the critical thinking skills we discussed in Chapter 6, it becomes clear that schools could offer comprehensive training in all of the thinking skills and dispositions that are essential for wiser reasoning.[20] In each case, evidence shows that these interventions will improve academic achievement for people of lower cognitive ability,[21] while also discouraging the dogmatic, closed-minded, and lazy thinking that often plagues people of high intelligence and expertise.

These benefits have been documented throughout the education system—from elementary schoolchildren to university undergraduates. But we can only cultivate wiser thinking if we allow students—even those just starting school—to face occasional moments of confusion and frustration rather than spoon-feeding them easy-to-digest information.

~

You don't need to be a teacher or a child to benefit from these findings. For work or for pleasure, most of us continue to learn into adulthood, and being able to regulate our own studying is essential if we are to make the most of our learning opportunities. The research shows that most people—even those of great intelligence—use poor learning techniques; the strategic use of desirable difficulties can improve your memory while also training your brain to be better equipped to deal with confusion and uncertainty in any context.[22]

You can:

- Space out your studies, using shorter chunks distributed over days and weeks. Like the postmen in Baddeley's initial experiment, your progress may feel slow compared with the initial head-start offered by more intensive study. But by forcing yourself to recall the material after the delay between each session, you will strengthen the memory trace and long-term recall.

- Beware of fluent material. As discussed previously, superficially simple textbooks can lead you to believe that you are learning well, while, in fact, they are reducing your long-term recall. So try to study more nuanced material that will require deeper thinking, even if it is initially confusing.
- Give yourself a pre-test. As soon as you begin exploring a topic, force yourself to explain as much as you already know. Even if your initial understanding is abysmally wrong, experiments show that this prepares the mind for deeper learning and better memory overall, as you correct for your errors in your subsequent studies.
- Vary your environment. If you tend to study in the same place for too long, cues from that environment become associated with the material, meaning that they can act as nonconscious prompts. By ensuring that you alter the places of learning, you avoid becoming too reliant on those cues—and like other desirable difficulties, this reduces your immediate performance but boosts your long-term memory. In one experiment, simply switching rooms during studying resulted in 21 percent better recall on a subsequent test.
- Learn by teaching. After studying—and without looking at your notes—imagine that you are explaining all that you have covered to another person. Abundant evidence shows that we learn best when we have to teach what we have just learned, because the act of explanation forces us to process the material more deeply.
- Test yourself regularly. So-called "retrieval practice" is by far the most powerful way of boosting your memory. But make sure you don't give in and look at the answers too quickly. The temptation is to look up the answer if it doesn't immediately come to mind, but you need to give yourself a bit of time to really struggle to recall, otherwise you won't be exercising your memory enough to improve long-term recall.
- Mix it up. When testing yourself, you should make sure you combine questions from different topics rather than only focusing on

one subject. Varying the topic forces your memory to work harder to recall the apparently unrelated facts, and it can also help you to see underlying patterns in what you are learning.

- Step outside your comfort zone and try to perform tasks that will be too difficult for your current level of expertise. And try to look for multiple solutions to a problem rather than a single answer. Even if none of your solutions is perfect, these productive failures will also increase your conceptual understanding.
- When you are wrong, try to explain the source of the confusion. Where did the misconception come from—and what was the source of the error? Not only does this prevent you from making the same specific error again; it also strengthens your memory of the topic as a whole.
- Beware the foresight bias. As Robert and Elizabeth Bjork have shown, we are bad at judging the level of our learning, based on our current performance—with some studies showing that the *more* confident we are in our memory of a fact, the *less* likely we are to remember it later. This, again, is down to fluency. We are more confident of things that initially come to mind easily—but we often haven't processed those fluent facts very deeply. So be sure to test yourself regularly on the material that you think you know well, in addition to the material that may feel less familiar.

Besides aiding factual learning, desirable difficulties can also help you to master motor skills, like playing a musical instrument. The current dogma is that music practice should be a disciplined but highly repetitive affair, where you spend lengthy sessions practicing the same few bars again and again until you play them to near perfection.

Instead, the Bjorks' research suggests that you would do far better to alternate a few different excerpts of music, spending a few minutes on each one. This will cause you to refresh your memory each time you come back to the exercise or excerpt.[23]

You can also try to inject some variability into the performance of the music itself. As part of her research on "conditional learning," Ellen Langer told a group of piano students to "change your style every few minutes and not to lock into one particular pattern. While you practice, attend to the context, which may include subtle variations or any feelings, sensations, or thoughts you are having." Independent judges considered their playing to be more proficient on a subsequent test than that of students who had instead practiced in a more traditional manner, aimed at rote memorization.

Langer has since replicated this experiment in a large symphony orchestra—where endless repetitive practice often leads to burnout. When asked to look for the subtle nuances in their performance, the players' enjoyment increased, and their performance was also judged to be more enjoyable by an independent group of musicians.[24]

The orchestra may seem to be a very different environment from the school classroom, but the philosophy of deliberately embracing nuance and complexity in learning can be applied to any context.

~

After having met the Bjorks at UCLA, I visited a nearby school in Long Beach, California, that may be the most comprehensive attempt yet to apply all the principles of evidence-based wisdom in a single institution.

The school is called the Intellectual Virtues Academy and it is the brainchild of Jason Baehr, a professor of philosophy at Loyola Marymount University in Los Angeles. Baehr's work focuses on "virtue epistemology"—examining the philosophical importance of character traits like intellectual humility, curiosity, and open-mindedness for good reasoning—and he has recently collaborated with some of the psychologists studying intellectual humility.

At the time of the IVA's founding, Baehr's interest had been purely theoretical, but that changed with a phone call from his friend and fellow philosopher Steve Porter, who heard a radio news program about the Obamas' choice of school for their daughters. The piece happened to mention the availability of so-called "charter" schools—state-funded schools that are managed privately according to their own educational vision and curriculum.

The two philosophers both had young children of their own, so why not attempt to set up their own charter school, Porter suggested. They began to meet up regularly in coffee shops to discuss how they might apply a model of teaching that deliberately cultivates intellectual virtues like curiosity—"not as an add-on, extra-curricular program, but rather where everything is oriented around the question of how we can help our students to grow in some of these qualities," Baehr told me.

This vision is clear from the moment I step into the building. Written on the walls of every classroom are the nine "master virtues" that the IVA considers to be crucial for good thinking and learning, with accompanying slogans. They are divided into three categories:

Getting started

- **Curiosity:** a disposition to wonder, ponder, and ask why. A thirst for understanding and a desire to explore.
- **Intellectual humility:** a willingness to own up to one's intellectual limitations and mistakes, unconcerned with intellectual status or prestige.
- **Intellectual autonomy:** a capacity for active, self-directed thinking. An ability to think and reason for oneself.

Executing well

- **Attentiveness:** a readiness to be "personally present" in the learning process. Keeps distractions at bay. Strives to be mindful and engaged.
- **Intellectual carefulness:** a disposition to notice and avoid intellectual pitfalls and mistakes. Strives for accuracy.
- **Intellectual thoroughness:** a disposition to seek and provide explanations. Unsatisfied with mere appearances or easy answers. Probes for deeper meaning and understanding.

Handling challenges

- **Open-mindedness:** an ability to think outside the box. Gives a fair and honest hearing to competing perspectives.
- **Intellectual courage:** a readiness to persist in thinking or communicating in the face of fear, including fear of embarrassment or failure.

- **Intellectual tenacity:** a willingness to embrace intellectual challenge and struggle. Keeps its "eyes on the prize" and doesn't give up.

As you can see, some of these virtues, including intellectual humility, open-mindedness, curiosity, are exactly the same elements of good thinking that Igor Grossmann had included in his initial studies of wise reasoning about everyday events, while others—such as "intellectual carefulness" and "intellectual thoroughness"—are more closely linked to the cultivation of skepticism that we explored in Chapter 6; "intellectual courage," "intellectual tenacity" and "intellectual autonomy," meanwhile, enforce the idea of struggle and confusion that Stigler and the Bjorks have studied so extensively.

Whether or not you are interested in the IVA's educational model, that's a pretty good checklist for the kinds of mental qualities that are essential for individuals to avoid the intelligence trap.

The children are taught about these concepts explicitly with a weekly "advisory" session, led by teachers and parents. During my visit, for instance, the advisory sessions explored "effective listening," in which the children were encouraged to reflect on the way they talk with others. They were asked to consider the benefits of some of the virtues, such as intellectual humility or curiosity, during conversation and, crucially, the kinds of situations where they may also be inappropriate. At the end of the session, the class listened to an episode from the *This American Life* podcast, about the real-life story of a young girl, Rosie, who struggled to communicate with her workaholic father, a physicist—a relatable exercise in perspective taking. The aim, in all this, is to get the children to be more analytical and reflective about their own thinking.

Besides these explicit lessons, the virtues are also incorporated into the teaching of traditional academic subjects. After the advisory session, for instance, I attended a seventh-grade (consisting of twelve- to thirteen-year-olds) lesson led by Cari Noble. The students there were learning about the ways to calculate the interior angles of the polygon, and rather than teaching the principles outright, the class had to struggle through the logic of coming up with the formula themselves—a strategy that reminded me a lot of Stigler's accounts of the Japanese classroom. Later on, I saw an English class in which the students discuss music appreciation, which included a TED talk by the conductor Benjamin Zander, in which he discusses his own difficulties with learning the piano—again promoting the idea that intellectual struggle is essential for progress.

Throughout the day, the teachers also "modeled" the virtues themselves, making sure to admit their own ignorance if they didn't immediately know an answer—an expression of intellectual humility—or their own curiosity if something suddenly led their interest in a new direction. As Dweck, Engel, and Langer have all shown, such subtle signals really can prime a child's own thinking.

I only visited the school for one day, but my impression from my conversations with the staff was that its strategy faithfully builds on robust psychological research to ensure that more sophisticated reasoning is incorporated into every subject, and that this doesn't sacrifice any of their academic rigor. As the principal, Jacquie Bryant, told me: "Students couldn't practice the intellectual virtues if they were not up against a challenging and complex curriculum. And we can't gauge the depth of their understanding unless we have the writing and feedback. The two go together."

The children's metacognition—their awareness of potential thinking errors and their capacity to correct them—appeared, from my observations, to be incredibly advanced compared to the average teenager.

Their parents certainly seem to be impressed. "This is great. A lot of us don't learn this until adulthood—if we ever learn it," Natasha Hunter, one of the parent advisors, told me during the advisory lesson. Hunter also teaches at a local college, and she was excited to see the children's sophisticated reasoning at such a young age. "I think that critical thinking has to happen at this level. Because when they get to me, after public school education, they aren't thinking at the level at which we need them to think."

The students' academic results speak for themselves. In its first year the IVA was scored as one of the top three schools in the Long

Beach Unified School District. And in a state-wide academic achievement test for the 2016–17 school year, more than 70 percent of its students achieved the expected standards in English, compared to a 50 percent average across California.[25]

We have to be wary of reading too much into this success. The IVA is only one school with highly motivated staff who are all dedicated to maintaining its vision, and many of the psychologists I've spoken to point out that it can be difficult to encourage effective, widespread educational reform.

Even so, the IVA offered me a taste of the way that Western education could begin to cultivate those other thinking styles that are so important for effective reasoning in adult life—producing a whole new generation of wiser thinkers.

PART 4

The folly and wisdom of the crowd: *How teams and organizations can avoid the intelligence trap*

9

The makings of a "dream team": *How to build a supergroup*

According to most pundits, Iceland should have had no place at the Euro 2016 men's soccer tournament. Just four years previously, they had ranked 131st in the world.[1] How could they ever hope to compete as one of the top twenty-four teams entering the championships?

The first shock came during the qualifying rounds in 2014 and 2015, when they knocked out the Netherlands to become the smallest nation ever to reach the championships. Then came their surprise draw with Portugal at the Saint-Etienne stadium in the first round. Their unexpected success was enough to rattle Portugal's star player, Cristiano Ronaldo, who sulked about the tactics. "It was a lucky night for them," he told reporters after the match. "When they don't try to play and just defend, defend, defend, this in my opinion shows a small mentality and they are not going to do anything in the competition."

The Icelanders were undeterred. They drew their next match with Hungary, and beat Austria 2–1. Commentators were sure that the tiny nation's luck would soon run out. Yet they did it again—this time against England, a team composed almost entirely of players from the world's top-twenty elite football clubs. The UK's TV commentators were rendered literally speechless by the final goal;[2] the *Guardian* described the match as one of the most "humiliating defeats in England's history."[3]

Iceland's dream ended when they finally succumbed to the home team in the quarter-finals, but football pundits from across the globe were nevertheless gobsmacked by their success. As one of *Time*'s sportswriters, Kim Wall, put it: "Iceland's very presence at the Championships came against all the odds. A volcanic island blanketed by year-round glaciers, the nation has the world's shortest soccer season: even the national stadium's designer pitch grass, handpicked to withstand Arctic winds and snow, occasionally froze to death."[4] And with a population of 330,000, they had a smaller pool of potential players than many London boroughs; one of their coaches still worked part time as a dentist.[5] In many people's eyes, they were the true heroes of the tournament, not Portugal, the ultimate winner.

At the time of writing (in 2018) Iceland remained at around twenty in the world rankings, and had become the smallest country ever to qualify for the World Cup; contrary to Ronaldo's criticisms, their success was not a mere fluke after all. How had this tiny nation managed to beat countries that were more than twenty times its size, their teams composed of some of the sport's greatest superstars?

Is it possible that their unexpected success came because—and not in spite—of the fact that they had so few star players?

~

The history of sport is full of surprising twists of fate. Perhaps the most famous upset of all is the "Miracle on Ice"—when a team of American college students beat the accomplished Soviet hockey team at the 1980 Winter Olympics. More recently, there was Argentina's surprise gold medal in basketball at the 2004 Olympics, where they thrashed the US—the firm favorites. In each case, the underdogs had less established players, yet somehow their combined talent was greater than the sum of their parts. But in terms of the sheer audacity of their challenge—and the teamwork that allowed them to punch so far above their weight—Iceland's success is perhaps the most instructive.

Sporting talent is very different from the kinds of intelligence we have been exploring so far, but the lessons from such unexpected successes may stretch beyond the football pitch. Many organizations employ highly intelligent, qualified people in the assumption that they will automatically combine their collective brainpower to produce magical results. Inexplicably, however, such groups often fail to cash in on their talents, with poor creativity, lost efficiency, and sometimes overly risky decision making.

In the past eight chapters, we have seen how greater intelligence and expertise can sometimes backfire for the individual, but the very same issues can also afflict teams, as certain traits, valued in high-performing individuals, may damage the group as a whole. You really can have "too much talent" in a team.

This is the intelligence trap of not one brain, but many, and the same dynamics that allowed Iceland to beat England can also help us to understand workplace politics in any organization.

~

Before we look specifically at the dynamics that link the England football team to the corporate boardroom, let's first consider some more general intuitions about group thinking.[6]

One popular idea has been the "wisdom of the crowd"—the idea that many brains, working together, can correct for each other's errors in judgments; we make each other better.* Some good evidence of this view comes from an analysis of scientists' journal articles, which finds that collaborative efforts are far more likely to be cited and applied than papers with just one author. Contrary to the notion of a lone genius, conversations and the exchange of ideas bring out the best in the team members; their combined brainpower allows them to see connections that had been invisible previously.[7]

Yet there are also plenty of notorious examples where team thinking fails, sometimes at great cost. Opposing voices like to point to the phenomenon of "groupthink," first described in detail by the Yale

---

* An argument for the wisdom of the crowds can also be traced back to Charles Darwin's cousin, Francis Galton. For an article published in *Nature* journal in 1907, he asked passersby at a country fair to estimate the weight of an ox. The median estimate came in at 1,198 lb—just 9 lb (or 0.8 percent) off the correct value. And more than 50 percent of the estimates were within around 4 percent either side of the true value. Building on this finding, some commentators have argued that reaching a group consensus is often the best way to increase the accuracy of our judgments—and you're probably going to be guaranteed greater success by recruiting as many talented individuals as possible.

University psychologist Irving Janis. Inspired by the Bay of Pigs disaster in 1961, he explored the reasons why the Kennedy administration decided to invade Cuba. He concluded that Kennedy's advisors had been too eager to reach a consensus decision and too anxious about questioning each other's judgments. Instead, they reinforced their existing biases—exacerbating each other's motivated reasoning. The individuals' intelligence didn't matter very much once the desire to conform had blinded their judgment.

Skeptics of collective reasoning may also point to the many times that groups simply fail to agree on any decision at all, reaching an impasse, or they may overly complicate a problem by incorporating all the points of view. This impasse is really the opposite of the more single-minded groupthink, but it can nonetheless be very damaging for a team's productivity. You want to avoid "design by committee."

The latest research helps us to reconcile all these views, offering some clever new tools to determine whether a group of talented people can tap into their combined ability or whether they will fall victim to groupthink.

~

Anita Williams Woolley has been at the forefront of these new findings, with the invention of a "collective intelligence" test that promises to revolutionise our understanding of group dynamics. I met her in her lab at Carnegie Mellon University in Pittsburgh, where she was conducting the latest round of experiments.

Designing the test was a Herculean task. One of the biggest challenges was designing a test that captured the full range of thinking that a group has to engage with: brainstorming, for instance, involves a kind of "divergent" thinking that is very different from the more restrained, critical thinking you may need to come to a decision. Her team eventually settled on a large battery of tasks—lasting five hours in total—that together tested four different kinds of thinking: *generating* new ideas; *choosing* a solution based on sound judgment; *negotiating* to reach compromise; and finally, general ability at *task execution* (such as coordinating movements and activities).

Unlike an individual intelligence test, many of the tasks were practical in nature. In a test of negotiation skills, for instance, the groups had to imagine that they were housemates sharing a car on a trip into town, each with a list of groceries—and they had to plan their trip to get the best bargains with the least driving time. In a test of moral reasoning, meanwhile, the subjects played the role of a jury, describing how they would judge a basketball player who had bribed his instructor. And to test their overall execution, the team members were each sat in front of a separate computer and asked to enter words into a shared online document—a deceptively simple challenge that tested how well they could coordinate their activities to avoid repeating words or writing over each other's contributions.[8] The participants were also asked to perform some verbal or abstract reasoning tasks that might be included in a traditional IQ test—but they answered as a group, rather than individually.

The first exciting finding was that each team's score on one of the constituent tasks correlated with its score on the other tasks. In other words, there appeared to be an underlying factor (rather like the "mental energy" that is meant to be reflected in our general intelligence) that meant that some teams consistently performed better than others.

Crucially, and in line with much of the work we have already seen on individual creativity, decision making and learning—a group's success appeared to only modestly reflect the members' average IQ (which could explain just 2.25 percent of the variation in collective intelligence). Nor could it be strongly linked to the highest IQ within the group (which accounted for 3.6 percent of the variation in collective intelligence). The teams weren't simply relying on the smartest member to do all the thinking.

Since they published that first paper in *Science* in 2010, Woolley's team has verified their test in many different contexts, showing that it can predict the success of many real-world projects. Some were conveniently close to home. They studied students completing a two-month group project in a university management course, for instance. Sure enough, the collective intelligence score predicted the team's performance on various assignments. Intriguingly, teams with a higher collective intelligence kept on building on their advantage during this project: not only were they better initially; they also improved the most over the eight weeks.

Woolley has also applied her test in the army, in a bank, in teams of computer programmers, and at a large financial services company, which ironically had one of the lowest collective intelligence scores she had ever come across. Disappointingly, she wasn't asked back; a symptom, perhaps, of their poor groupthink.

The test is much more than a diagnostic tool, however. It has also allowed Woolley to investigate the underlying reasons why some teams have higher or lower collective intelligence—and the ways those dynamics might be improved.

One of the strongest and most consistent predictors is the team members' social sensitivity. To measure this quality, Woolley used a classic measure of emotional perception, in which participants are given photos of an actor's eyes and asked to determine what emotion that person is supposed to be feeling—whether they are happy, sad, angry or scared, with the participants' average score strongly predicting how well they would perform on the group tasks. Remarkably, the very same dynamics can determine the fate of teams working together remotely, across the internet.[9] Even though they aren't meeting face to face, greater social sensitivity still allows them to read between the lines of direct messages and better coordinate their actions.

Beyond the "reading the mind in the eyes" test, Woolley has also probed the specific interactions that can elevate or destroy a team's thinking. Companies may value someone who is willing to take charge when a group lacks a hierarchy, for instance—the kind of person who may think of themselves as a "natural leader." Yet when Woolley's team measured how often each member spoke, they found that the better groups tend to allow each member to participate equally; the worst groups, in contrast, tended to be dominated by just one or two people.

Those more domineering people don't have to be excessively loud or rude, but if they give the impression that they know everything already, other team members will feel they have nothing to contribute, which deprives the group of valuable information and alternative points of view.[10] Untempered enthusiasm can be a vice.

The most destructive dynamic, Woolley has found, is when team members start competing against each other. This was the problem with the financial services company and their broader corporate culture. Each year, the company would only promote a fixed number of individuals based on their performance reviews—meaning that each employee would feel threatened by the others, and group work suffered as a result.

Since Woolley published those first results, her research has garnered particular interest for its insights into sexism in the workplace. The irritating habits of some men to "mansplain," interrupt and appropriate women's ideas has been noted by many commentators in recent years. By shutting down a conversation and preventing women from sharing their knowledge, those are exactly the kinds of behaviors that sabotage group performance.

Sure enough, Woolley has shown that—at least in her experiments in the US—teams with a greater proportion of women have a higher collective intelligence, and that this can be linked to their higher overall social sensitivity, compared to groups consisting of a larger proportion of men.[11] This was equally true when Woolley tested the collective intelligence of online teams playing the *League of Legends* computer game, when the players' gender was obscured by their avatar: it wasn't simply that the men were acting differently when they knew a woman was present.[12]

We don't yet know the exact cause of these gender differences. There may be a biological basis—testosterone has known effects on behavior and higher levels make people more impulsive and dominant, for instance—but some differences in social sensitivity could be culturally learned too.

Woolley told me that these findings have already changed opinions. "Some organizations have taken what we have found and just flipped it into hiring more women."

Whether or not you would deliberately change the gender balance on the basis of these findings, hiring people of both sexes with greater social sensitivity is an obvious way to boost the collective intelligence of an organization.

The very name—soft skills—that we attribute to social intelligence often implies that it is the weaker, secondary counterpart to other forms of intelligence, and the tests we use to explore interpersonal dynamics—such as the Myers-Briggs Type Inventory—are poor predictors of actual behavior.[13] If you are attempting to recruit a smart team, Woolley's research strongly suggests that these social skills should be a primary concern, and in the same way that we measure cognitive ability using standardized tests, we should begin to use scientifically verified measures to assess this quality.

~

By showing that collective intelligence has such a weak correlation with IQ, Woolley's tests begin to explain why some groups of intelligent people fail. But given the research on the individual intelligence trap, I was also interested in whether high performers are ever at an even *higher* risk of failing than teams of less average ability.

Intuitively, we might suspect that especially smart or powerful people will struggle to get along, due to the overconfidence or closed-mindedness that comes with their status, and that this might damage their overall performance. But are these intuitions justified?

Angus Hildreth at Cornell University can offer us some answers. His research was inspired by his own experiences at a global consulting firm, where he often oversaw meetings from some of the top executives. "They were really effective individuals who had got to where they were because they were good at what they did, but when they came together in these group contexts I was surprised at the dysfunction and the difficulties they faced," he told me during one of his frequent visits back home to London. "I was expecting this platonic ideal of leadership: you put all the best people in the room and obviously something good is going to happen. But there was an inability to make decisions. We were always running behind."

Returning to study for a PhD in organizational behavior at the University of California, Berkeley, he decided to probe the phenomenon further.

In one experiment, published in 2016, he gathered executives from a multinational healthcare company, assigned them to groups and asked them to imagine that they were recruiting a new chief financial officer from a selection of dummy candidates. The spread of power between the groups was not equal, however: some were composed of high-flying executives, who managed lots of people, while others were mostly made up of their subordinates. To ensure that he was not just seeing the effects of existing competition between the executives, he ensured that the executives within the groups had not worked together previously. "Otherwise there might be this past history, where someone had beaten someone else to a position."

Despite their credentials and experience, the groups of high-flyers often failed to reach a consensus; 64 percent of the high-power teams reached an impasse, compared with just 15 percent of the low-power teams; that's a four-fold difference in the groups' effectiveness.[14]

One problem was "status conflict"—the high-power members were less focused on the task itself, and more interested in asserting their authority in the group and determining who would become top dog. But the high-flying teams were also less likely to share information and integrate each other's points of view, making it much harder to come to a successful compromise.

You could argue that you need greater confidence to get ahead in the first place; perhaps these people had always been a little more self-serving. But a further lab experiment, using students as participants, demonstrated that it takes surprisingly little effort to send people on that kind of ego trip.

The students' first task was simple: they were separated into pairs and told to make a tower using building blocks. In each pair, one was told that they were a leader and the other a follower, ostensibly based on their answers to a questionnaire. Their success or failure in the task was unimportant; Hildreth's goal was to prime some of the participants with a sense of power. In the next exercise, he rearranged the students into groups of three, composed either of all leaders or all followers, and set them some tests on creativity, such as inventing a new organization and laying out its business plan.

Drunk on the tiny bit of power bestowed in the previous exercise, the former leaders tended to be less cooperative and found it harder to share information and agree on a solution, dragging the groups' overall performance down. They were demonstrating exactly the kind of mutually sabotaging behaviors, in other words, that Woolley had found to be so destructive to a team's collective intelligence.

Hildreth says the power struggles were plain to see as he observed the groups at work. "They were pretty cold interactions," he told me. "Reasonably often at least one student in the group withdrew, because the dynamics were so uncomfortable—or they just didn't want to engage in the conversation because their ideas weren't being heard. They thought, 'I'm the one who makes the decisions, and my decisions are the best ones.'"

Although Hildreth's study only explored these dynamics in a single healthcare company, various field studies suggest these dynamics are ubiquitous. An analysis of Dutch telecommunications and financial institutions, for instance, examined behavior in teams across the company's hierarchies, finding that the higher up the company you go, the greater the level of conflict reported by the employees.

Crucially, this seemed to depend on the members' own understanding of their positions in the pecking order. If the team, as a whole, agreed on their relative positions, they were more productive, since they avoided constant jockeying for authority.[15] The worst groups were composed of high-status individuals who didn't know their rank in the pecking order.

The most striking example of these power plays—and the clearest evidence that too much talent can be counterproductive—comes from a study of "star" equity analysts in Wall Street banks. Each year, *Institutional Investor* ranks the top analysts in each sector, offering them a kind of rock star status among their colleagues that can translate to millions of dollars of increased earnings; they are also regularly picked as media pundits. Needless to say, these people often flock together at the same prestigious firms, but that doesn't always bring the rewards the company might have hoped.

Studying five years of data across the industry, Boris Groysberg of the Harvard Business School found that teams with more star players do indeed perform better, but only up to a certain point, after which the benefits of additional star talent tailed off. And with more than 45 percent of the department filled with *Institutional Investor*'s picks, the research department actually becomes *less* effective.

The groups appeared to be particularly fragile when the stars' areas of expertise happened to coincide, putting them in more direct competition with each other, and it was less of a factor when they fell into different sectors, and were therefore in less direct competition with each other. Then, the company could afford to recruit a few more stars—up to around 70 percent of the workforce—before their rutting egos destroyed the team's performance.[16]

~

Hildreth's theory is based on the group interactions among the powerful. But besides disrupting communication and cooperation, status conflict can also interfere with the brain's information processing ability. At least for the duration of the meeting, the individual members can themselves be a little bit more stupid as a result of their interactions.

The study, which took place at Virginia Tech, gathered small groups of people and gave them each some abstract problems, while broadcasting their progress—relative to the other team members—on their computer interface. The feedback turned out to paralyze some of the candidates, lowering their scores compared to their performance on a previous test. Despite having started out with roughly equal IQs, the participants eventually separated into two distinct strata, with some people appearing to be particularly sensitive to the competition.[17]

The diminished brainpower was also evident in fMRI scans taken at the time of the test: it appeared to be associated with increased brain activity in the amygdala—an almond-shaped bundle of neurons, deep in the brain, associated with emotional processing—and reduced activity in prefrontal cortices behind the forehead, which are associated with problem solving.

The team concluded that we can't separate our cognitive abilities from the social environment: all the time, our capacity to apply our brainpower will be influenced by our perceptions of those around us.[18] Given these findings, it's easy to see how the presence of a brilliant—but arrogant—team member may hurt both the collective and individual intelligence of his or her more sensitive colleagues, a double whammy that will reduce their productivity across the board.

As one of the researchers, Read Montague, puts it: "You may joke about how committee meetings make you feel brain dead, but our findings suggest that they may make you act brain dead as well."

~

The sports field may seem a far cry from the boardroom, but we see exactly the same dynamics in many sports.

Consider the fate of the Miami Heat basketball team in the early 2010s. After signing up LeBron James, Chris Bosh, and Dwayne Wade—the "Big Three"—the team was overflowing with natural talent—but they ended the 2010–11 season ranked twenty-ninth out of thirty teams. It was only after Bosh and Wade fell out of the game with injuries that they would eventually win an NBA championship the following year. As the sportswriter Bill Simmons put it: "Less talent became more."[19]

To find out if this is a common phenomenon, the social psychologist Adam Galinsky first examined the performance of football (soccer) teams in the 2010 World Cup in South Africa and the 2014 World Cup in Brazil. To determine the country's "top talent," they calculated how many of its squad were currently on the payroll of one of the top thirty highest-earning clubs listed in the Deloitte Football Money League (which includes Real Madrid, FC Barcelona, and Manchester United). They then compared this value to the country's ranking in the qualifying rounds.

Just as Groysberg had observed with the Wall Street analysts, Galinsky's team found a "curvilinear" relationship; a team benefited from having a few stars, but the balance seemed to tip at about 60 percent, after which the team's performance suffered.

The Dutch football team offered a perfect case in point. After disappointing results in the Euro 2012 championships, the coach, Louis van Gaal, reassembled the team—reducing the percentage of "top talent" from 73 percent to 43 percent. It was an extraordinary move, but it seems that he had judged the dynamics correctly: as Galinsky and his coauthors point out in their paper, the Netherlands did not lose a single game in the qualifying rounds of the 2014 World Cup.

The too-much-talent effect in football

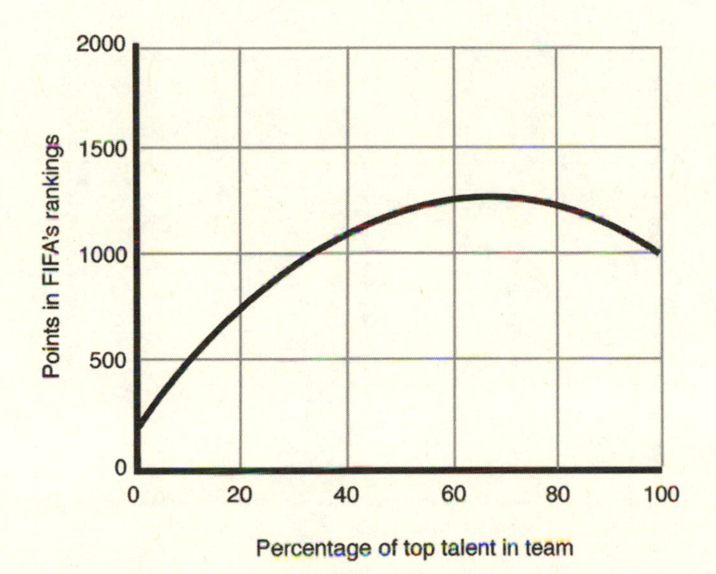

Source: Swaab, R. I., Schaerer, M., Anicich, E. M., Ronay, R., & Galinsky, A. D. (2014). The too-much-talent effect: Team interdependence determines when more talent is too much or not enough. *Psychological Science, 25*(8), 1581–1591.

To check the "too-much-talent" effect in a new context, Galinsky then applied the same thinking to basketball rankings, looking at the ten NBA seasons from 2002 to 2012. To identify the star players, they used a measure of "estimated wins added"—which uses the statistics of the game to calculate whether a team member was often a deciding factor in a match's outcome. The "top talent," Galinsky's team decided, lay in the top third of these rankings—an admittedly arbitrary cutoff point, but one that is often used in many organizations to decide exceptional performance. Crucially, many of the players within their ranking coincided with the players selected for NBA's own All-Star Game, suggesting it was a valid measure of top talent.

Once again, the researchers calculated the proportion of star players within each club, and compared it to the team's overall wins within each season. The pattern was almost identical to the results from the football World Cup:

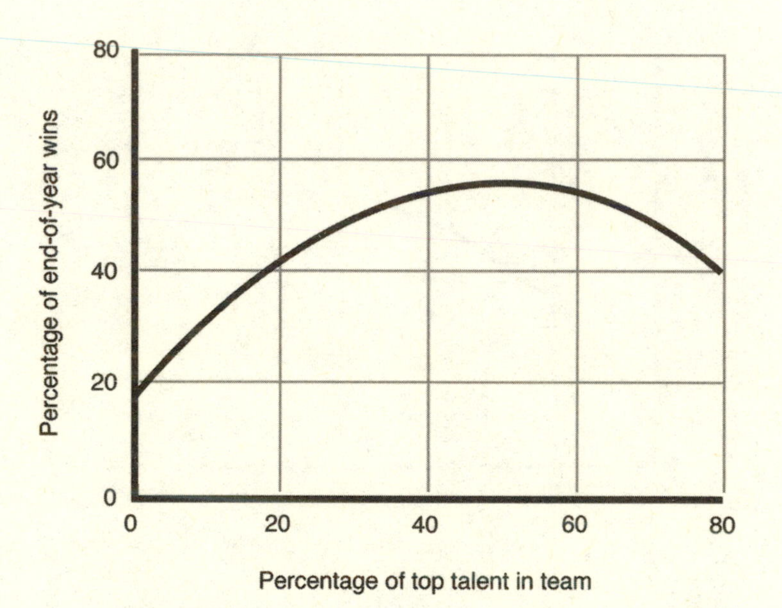

Source: Swaab, R. I., Schaerer, M., Anicich, E. M., Ronay, R., & Galinsky, A. D. (2014). The too-much-talent effect: Team interdependence determines when more talent is too much or not enough. *Psychological Science, 25*(8), 1581–1591.

In one final experiment, the team examined data from Major League Baseball, a sport that does not require so much coordination between players. Here, they found no evidence of the too-much-talent effect, which supports the idea that status is only harmful when we need to cooperate and bring out the best in each other.[20] For sports, like baseball, that are less interdependent than basketball or football, it pays to buy all the top talent you can afford.

If we look back at Iceland's unexpected victory against England in the Euro 2016 football championships, it's clear that its success stems from many different factors. The country had spent years investing in improving its training programs, and the team was in excellent hands under Swedish coach Lars Lagerbac and his assistant Heimir Hallgrímsson. The quality of the individual players was undoubtedly better than it ever had been. But although many worked for international football clubs, just one of them at the time (Gylfi Sigurðsson) had a contract in one of the top-thirty clubs in Deloitte's Football Money League. They hadn't yet achieved the kind of international status that could be so destructive.

England, in contrast, had pulled twenty-one of its twenty-three players from these super-rich teams, meaning they accounted for more than 90 percent of the squad, far above the optimum threshold. In fact, according to my own calculations, not one of the teams that succeeded in passing through to the quarter-finals had as many star players (the closest was Germany, who had 74 percent). England's defeat against Iceland at the Allianz Riviera Stadium in Nice is an almost perfect fit with Galinsky's model.

Although they may not have been aware of Galinsky's scientific work, football pundits noted the disastrous team dynamics at the time of the tournament. "England, for all their individual talents, lacked so much as one," sports writer Ian Herbert wrote in the *Independent* after Iceland's win. "The reason why the nation struggles to feel empathy or connections with many of these players is the ego. Too famous, too important, too rich, too high and mighty to discover the pace and the fight and the new dimensions to put it on when against one of Europe's most diminutive football nations. That is this England."[21] The eventual champions, Portugal, incidentally took only four of their players from the elite clubs in the Deloitte Money League. The team may have had Cristiano Ronaldo—arguably the biggest star in the game—but it had not exceeded Galinsky's threshold.

The "Miracle on Ice" during the 1980 Winter Olympics at Lake Placid, New York, shows exactly the same pattern. The Soviet team had an unbroken record, returning home with gold in each of the past four Games. Of twenty-nine matches, they had won twenty-seven. Eight of their players had competed in at least some of those Olympic Games and also played for high-profile teams in their home country. The US team, in contrast, were a bunch of college kids, with an average age of twenty-one—making them the youngest team of the whole tournament—and little international experience.

The coach, Herb Brooks, was under no illusions that it was a case of "David against Goliath." Yet David triumphed; America beat the Soviet Union 4–3, and they entered the second medal round with Finland. They walked away with the gold medal.

And you don't need to be an international superstar for this kind of dynamic to apply to you and your colleagues; as Hildreth found with his university experiments, elite performance depends in part on your *perception* of your talents relative to those around you.

Anita Williams Woolley, who first alerted me to Galinsky's research, even saw it in her sons' amateur soccer team. "They had a very good team last year and won the state cup," she told me. "And then they attracted all of these really good players from other clubs and it ruined their dynamic. They've now lost five games this year."

~

With this new understanding of collective intelligence and the too-much-talent effect, we are now very close to being able to discover some simple strategies to improve *any* team's performance. Before we do so, however, we need to explore the role of the leader in more detail—and the best case study emerges from a tragedy on Everest's slopes.

On May 9, 1996, at an altitude of 26,000 feet, two expeditions were ready to depart from Camp IV on the South Col route in Nepal. Rob Hall, a thirty-five-year-old New Zealander, led the team from Adventure Consultants. They were joined by a group from Mountain Madness, led by forty-year-old Scott Fischer from Michigan. Each team included the leader, two additional guides, eight clients, and numerous Sherpas.

Their expertise was unquestionable. Hall had already reached the summit four times previously, successfully guiding thirty-nine clients in the process; he was known to be meticulous with his organization. Fischer had only mastered Everest once before, but he had earned his stripes on many of the world's most challenging peaks.

And he was confident of his methods. "We've got the Big E figured out . . . we've built a yellow brick road to the summit," one of the surviving members, Jon Krakauer, recalled Fischer saying.

Although they came from different companies, Hall and Fischer had decided to work together on the final bid—but they were soon beset with delays and difficulties. One of their Sherpas had failed to install a "fixed line" to guide their ascent—adding an hour to their journey—and a bottleneck of climbers began to amass further down the slope as they waited for the line to be installed. By early afternoon, it became clear that many would not be able to reach the summit and return by the time darkness fell. A few decided to turn back, but the majority—including Hall and Fischer—pressed on.

It proved to be a fatal decision. At 15:00, snow began to fall; by 17:30 it was a full-blown blizzard. Hall, Fischer, and three of their team members died on their descent.

Why did they decide to continue up to the summit, even when the conditions were worsening? Fischer himself had previously spoken of the "two o'clock rule"—to turn around if they failed to reach the summit by 2 p.m. so that they could be sure they were back at camp by nightfall. Yet they continued to their deaths.

Michael Roberto, from Harvard Business School, has analyzed accounts of the disaster—including a best-selling book by Jon Krakauer—and he believes that the teams' decisions may have been influenced by the now familiar cognitive biases, including the sunk cost error (the climbers had each invested $70,000 in the attempt, and weeks of potentially wasted effort) and overconfidence on the part of Hall and Fischer.[22]

More interesting for our purposes, however, were the group dynamics—and in particular, the hierarchy that Hall and Fisher had established around them. We have already seen some reasons why a hierarchy might be productive, since it could set to rest status conflict and infighting within a group.

In this case, however, the hierarchy backfired. Besides Hall and Fischer, there were more junior guides and local Sherpas with intimate knowledge of the peak, who might have corrected their errors. But the group did not feel comfortable expressing their concerns.

Krakauer describes a kind of strict "pecking order" that meant the clients were scared to question the guides, and the guides were scared to question the leaders, Hall and Fischer. As one of the more junior guides, Neal Beidleman, later put it: "I was definitely considered the third guide . . . so I tried not to be too pushy. As a consequence, I didn't always speak up when maybe I should have, and now I kick myself for it." Another guide, Anatoli Boukreev, was similarly wary of expressing his concerns that the team had not acclimatised to the thin air. "I tried not to be argumentative, choosing instead to downplay my intuitions."

According to Krakauer, Hall had made his own feelings on the hierarchy quite clear before they departed: "I will tolerate no dissension up there. My word will be absolute law, beyond appeal."

Lou Kasischke, one of the team who had decided to turn back, agreed. "You need candor between the leader and the followers," he told PBS. On an expedition, he said, the leader needs feedback from his or her team, but Hall wasn't receptive to those opinions. "Rob didn't foster that relationship where he expected us to say these things."[23] A hierarchy, then, can be both a productive and a dangerous thing.

We have to be wary of basing our conclusions on a single case-study, but Adam Galinsky has confirmed this conclusion by analyzing records of 5,104 Himalayan expeditions. Unable to question all the climbers themselves, he instead examined cultural differences in attitudes to authority. Various studies have demonstrated that some nationalities are more likely to believe you should strictly respect people's position within a group, while others accept that you can challenge and question your superiors. People from China, Iran, and Thailand, for instance, tend to respect hierarchy far more than citizens of the Netherlands, Germany, Italy, or Norway, according to one widely accepted measure; the US, Australia, and the UK fall somewhere in the middle.[24]

Comparing this data with the Everest records, Galinsky found that teams composed of people from the countries that respected a hierarchy were indeed more likely to reach the summit—confirming the suspicions that a hierarchy boosts productivity and eases coordination between the team members. But crucially, they were also more likely to lose team members in the attempt.

To check that they weren't accidentally measuring other traits—such as individual determination—that might also have correlated with the prevailing cultural attitudes to hierarchy, and influenced their chances of success, Galinsky's group examined data from more than a thousand solo trips. Now, they found no overarching differences between the different cultures. It was their group interactions that really made a difference.[25]

The same dynamic may lie behind many business disasters. Executives at Enron, for instance, had a kind of reverence for those above, and disagreement or doubt was seen as a severe sign of disloyalty. To survive, they said, "you had to keep drinking the Enron water."

There is an apparent paradox in these findings: if team members clearly understand their place in the pecking order, overall group performance will be boosted; but this is true *only if* team members themselves feel that their opinions are valued, and that they can challenge their leaders in the event of problems arising or poor decisions being taken.

~

Stretching from the Saint-Etienne football stadium to Wall Street and the slopes of Everest, we've now seen how some common dynamics shape group interactions and determine a team's collective intelligence. This new research appears to have captured the forces underlying teamwork in any context.

And in precisely the same way that our understanding of the individual intelligence trap offers simple strategies to escape error, this research also suggests some tried-and-tested ways to avoid the most common mistakes in group reasoning.

From Woolley and Galinsky's research, we can change the way we recruit new team members. In light of the too-much-talent effect, it would be tempting to argue that you should simply stop selecting people of exceptional ability—particularly if your team's composition has already passed that magic threshold of 50–60 percent being "star" players.

At this stage it's probably best not to be too fixated on that number—the specific ratio will almost certainly depend on the personalities within a group and the amount of cooperation it requires—but the scientific research at least suggests that we need to place a greater emphasis on interpersonal skills that will enhance the team's collective intelligence, even if that means rejecting someone who scores far higher on more standard measures of ability. That may include judging someone's emotional perceptivity and communication skills—whether they draw people out and listen, or whether they have a tendency to interrupt and dominate. If you are leading a multinational team, you might also choose someone with high cultural intelligence (which we explored in Chapter 1) since they would find it easier to navigate the different social norms.[26]

Given what we know about status conflict, we can also improve the interactions of the talent you do have. Hildreth, for instance, found strategies to avoid the clash of egos during his previous job at a global consulting firm. One example, he says, is to underline each person's expertise at each meeting and their reason for appearing at the group, which helps ensure that they have the chance to share relevant experience. "Often that's kind of lost in the melee of conflict."

Hildreth also recommends allotting a fixed amount of time for each person to contribute his or her opinion at the start of the meeting. The topic of discussion need not be related to the problem at hand, but this practice allows each person to feel that he or she has already made a contribution to the group's functioning, further defusing the status conflict and easing the ensuing conversation. "You get a lot more equality within the discussion so that everyone contributes," Hildreth said. And when you finally come to the problem at hand, he suggests that you set out a firm strategy for when and how you will make the decision—whether it will be by unanimous or majority vote, for instance—to avoid the kind of impasse that may come when too many intelligent and experienced people butt heads.

Lastly, and most importantly, the leader should embody the kinds of qualities he or she wants to see in a team—and should be particularly keen to encourage disagreement.

It is here that the research of group thinking comes closest to the new science of evidence-based wisdom, as more and more organizational psychologists are coming to see how the intellectual humility of a leader not only improves their individual decision making but also brings knock-on benefits for their closest colleagues.

Using staff questionnaires to explore the top management teams of 105 technology companies, Amy Yi Ou at the National University of Singapore has shown that employees under a humble leader of this kind are themselves more likely to share information, collaborate in times of stress and contribute to a shared vision. By tapping into the collective intelligence, such businesses were better able to overcome challenges and uncertainty, ultimately resulting in greater annual profits a year later.[27]

Unfortunately, Ou says that CEOs themselves tend to be very split in their opinions on the virtue of humility, with many believing that it can undermine their team's confidence in their abilities to lead. This was true even in China, she says, where she had expected to see greater respect for a humble mind-set. "Even there, when I'm talking to those high-profile CEOs, they reject the term humility," she told me. "They think that if I'm humble, I can't manage my team well. But my study shows that it actually works."

History offers us some striking examples of these dynamics at play. Abraham Lincoln's capacity to listen to the dissenting voices in his cabinet—a "team of rivals"—is famously thought to have been one of the reasons that he won the Civil War—and it apparently inspired Barack Obama's leadership strategy as president.

Jungkiu Choi, head of consumer banking at Standard Chartered Banking in China, meanwhile, gives us one modern case study of humility at the top. Before he took the role, top executives had expected to receive the red-carpet treatment when they visited individual branches, but one of Jungkiu's first moves was to ensure that each meeting was far more informal. He would turn up unannounced and organize friendly "huddles" with the employees to ask how he could improve the business.

He soon found that these meetings generated some of the company's most fruitful ideas. One of his groups, for instance, had suggested that the bank change its operating hours, including weekend shifts, to match other shops in the area. Within months, they were earning more from those few hours than they had in the whole of the rest of the week. With every employee able to contribute to the bank's strategy, its entire service was transformed—and customer satisfaction rose by more than 50 percent within two years.[28]

We can also see this philosophy in Google's CEO Sundar Pichai, who argues that the leader's single role is to "let others succeed." As he explained in a speech to his alma mater, the Indian Institute of Technology-Kharagpur: "[Leadership is] less about trying to be successful (yourself), and more about making sure you have good people, and your work is to remove that barrier, remove roadblocks for them so that they can be successful in what they do."

Like the many other principles of good teamwork, the humility of the leader can bring benefits to the sports field. One study found that the most successful high-school basketball teams were those whose coaches saw themselves as a "servant" to the team, compared with those whose coaches saw themselves as sitting apart from and above their students.[29] Under the humbler coaches, the team players were more determined, better able to cope with failure, and won more games per season. The humility modeled by the coach pushed everyone to work a little harder and to support their other teammates.

Consider John Wooden, commonly regarded as the most successful college basketball coach of all time. He led UCLA to win ten national championships in twelve years, and between 1971 and 1974, they went undefeated for eighty-eight games. Despite these successes, Wooden's every gesture made it clear that he was not above the players on his team, as seen in the fact that he would help sweep the locker room after every game.

In the memoir *Coach Wooden and Me*, his former player and lifelong friend, Kareem Abdul-Jabbar, described many instances of Wooden's unfailing humility, even when dealing with difficult confrontations with his players. "It was mathematically inevitable that Coach would take to heart what one of his players said, feel compelled to patch things up, and teach us all a lesson in humility at the same time."[30] Wooden made it clear that they could learn from each other, him included—and the team went from strength to strength as a consequence.

~

After Iceland's unexpected success at the Euro 2016 tournament, many commentators highlighted the down-to-earth attitude of Heimir Hallgrímsson, one of the team's two coaches, who still worked part time as a dentist despite leading the national team. He was apparently devoted to listening and understanding others' points of view, and he tried to cultivate that attitude in all of his players.

"Team-building is a must for a country like ours; we can only beat the big teams by working as one," he told the sports channel ESPN. "If you look at our team, we have guys like Gylfi Sigurðsson at Swansea [Football Club], who is probably our highest-profile player, but he's the hardest worker on the pitch. If that guy works the hardest, who in the team can be lazy?"[31]

As with the other elements of evidence-based wisdom, the study of collective intelligence is still a young discipline, but by applying these principles you can help to ensure that your team members play a little more like Iceland, and a little less like England—a strategy that will allow each person to bring out the best in those around them.

10

Stupidity spreading like wildfire: *Why disasters occur—and how to stop them*

We are on an oil rig in the middle of the ocean. It is a quiet evening with a light breeze.

The team of engineers has finished drilling, and they are now trying to seal their well with cement. They have checked the pressure at the seal, and all seems to be going well. Soon extraction can begin, and the dollars will start rolling in. It should be time to celebrate.

But the pressure tests were wrong; the cement has not set and the seal at the bottom of the well is not secure. As the engineers happily sign off their job, oil and gas has started to build up within the pipe—and it's rising fast. In the middle of the engineers' celebrations, mud and oil starts spewing onto the rig floor; the crew can taste the gas on their tongues. If they don't act quickly, they will soon face a full-on "blowout."

If you have even a passing knowledge of the world news in 2010, you may think you know what happens next: an almighty explosion and the largest oil spill in history.

But in this case, it doesn't happen. Maybe the leak is far enough away from the engine room, or the wind is blowing, creating a movement of air that prevents the escaping gas from catching light. Or maybe the team on the ground simply notice the build-up of pressure and are able to deploy the "blowout preventer" in time. Whatever the specific reason, a disaster is averted. The company loses a few days of extraction—and a few million dollars of profits—but no one dies.

~

This is not a hypothetical scenario or a wishful reimagining of the past. There had been literally dozens of minor blowouts in the Gulf of Mexico alone in the twenty years before the Deepwater Horizon spill at the Macondo well in April 2010—but thanks to random circumstances such as the direction and speed of the wind, full-blown disasters never took place, and the oil companies could contain the damage.[1]

Transocean, the company in charge of cementing the Deepwater Horizon rig, had even experienced a remarkably similar incident in the North Sea just four months previously, when the engineers had also misinterpreted a series of "negative pressure tests"—missing signs that the seal of the well was broken. But they had been able to contain the damage before an explosion occurred, resulting in a few days' lost work rather than an environmental catastrophe.[2]

On April 20, 2010, however, there was no wind to dissipate the oil and gas, and thanks to faulty equipment, all the team's attempts to contain the blowout failed. As the escaping gas built up in the engine rooms, it eventually ignited, unleashing a series of fireballs that ripped through the rig.

The rest is history. Eleven workers lost their lives, and over the next few months, more than 200 million gallons of oil were released into the Gulf of Mexico, making it the worst environmental catastrophe in American history. BP had to pay more than $65 billion in compensation.[3]

Why would so many people miss so many warning signs? From previous near misses to a failed reading of the internal pressure on the day of the explosion, employees seemed to have been oblivious to the potential for disaster.

As Sean Grimsley, a lawyer for a US Presidential Commission investigating the disaster, concluded: "The well was flowing. Hydrocarbons were leaking, but for whatever reason the crew after three hours that night decided it was a good negative pressure test. . . . The question is why these experienced men out on that rig talked themselves into believing that this was a good test. . . . None of these men wanted to die."[4]

Disasters like the Deepwater Horizon explosion require us to expand our focus, beyond groups and teams, to the surprising ways that certain corporate cultures can exacerbate individual thinking errors and subtly inhibit wiser reasoning. It is almost as if the organization as a whole is suffering from a collective bias blind spot.

The same dynamics underlie many of the worst manmade catastrophes in recent history, from NASA's *Columbia* disaster to the *Concorde* crash in 2000.

You don't need to lead a multinational organization to benefit from this research; it includes eye-opening findings for anyone in employment. If you've ever worried that your own work environment is dulling your mind, these discoveries will help explain your experiences, and offer tips for the best ways to protect yourself from mindlessly imitating the mistakes of those around you.

~

Before we examine large-scale catastrophes, let's begin with a study of "functional stupidity" in the general workplace. The concept is the brainchild of Mats Alvesson at Lund University in Sweden, and André Spicer at the Cass Business School in London, who coined the term to describe the counter-intuitive reasons that some companies may actively discourage their employees from thinking.

Spicer told me that his interest stems from his PhD at the University of Melbourne, during which time he studied decision making at the Australian Broadcasting Corporation (ABC).[5] "They would introduce these crazy change management programs, which would often result in nothing changing except creating a huge amount of uncertainty."

Many employees acknowledged the flaws in the corporation's decision making. "You found a lot of very smart people thrown together in an organization and many of them would spend a lot of time complaining how stupid the organization was," Spicer told me. What really surprised him, however, was the number of people who failed to acknowledge the futility of what they were doing. "These extremely high-skilled and knowledgeable professionals were getting sucked into these crazy things, saying 'this is intelligent, this is rational,' then wasting an incredible amount of time."*

Years later he would discuss such organizational failings with Alvesson at a formal academic dinner. In their resulting studies, the pair of researchers examined dozens of other examples of organizational stupidity, from the armed forces to IT analysts, newspaper publishers and their own respective universities, to examine whether many institutions really do make the most of their staff's brains.

Their conclusions were deeply depressing. As Alvesson and Spicer wrote in their book, *The Stupidity Paradox*: "Our governments spend billions on trying to create knowledge economies, our firms brag about their superior intelligence, and individuals spend decades of their lives building up fine CVs. Yet all this collective intellect does not seem to be reflected in the many organizations we studied. . . . Far from being 'knowledge-intensive,' many of our most well-known chief organizations have become engines of stupidity."[6]

In parallel with the kinds of biases and errors behind the intelligence trap, Spicer and Alvesson define "stupidity" as a form of narrow thinking lacking three important qualities: reflection about basic underlying assumptions, curiosity about the purpose of your actions, and a consideration of the wider, long-term consequences of your behaviors.[7] For many varied reasons, employees simply aren't being encouraged to *think*.

This stupidity is often *functional*, they say, because it can come with some benefits. Individuals may prefer to go with the flow in the workplace to save effort and anxiety, particularly if we know there will be incentives or even a promotion in this for us later. Such "strategic ignorance" is now well studied in psychological experiments where participants must compete for money: often

---

* The same culture can also be seen in the BBC's offices—a fact the broadcaster itself lampoons in its mockumentary TV series *W1A*. Having worked at the BBC while researching this book, it occurs to me that deciding to create a three-series sitcom about your own organizational failings is perhaps *the definition* of functional stupidity.

participants choose not to know how their decisions affect the other players.[8] By remaining in the dark, the player gains some "moral wiggle room" (the scientific term) that allows them to act in a more selfish way.

We might also be persuaded by social pressure: no one likes a trouble-maker, after all, who delays meetings with endless questions. Unless we are actively encouraged to share our views, staying quiet and nodding along with the people around us can improve our individual prospects—even if that means temporarily turning off our critical capacities.

Besides helping the individual, this kind of narrow-minded, unquestioning approach can also bring some immediate benefits for the organization, increasing productivity and efficiency in the short term without the employees wasting time questioning the wisdom of their behaviors. The result is that some companies may—either accidentally or deliberately—actually encourage functional stupidity within their offices.

Spicer and Alvesson argue that many work practices and structures contribute to an organization's functional stupidity, including excessive specialization and division of responsibilities. A human resources manager may now have the very particular, single task of organizing personality tests, for instance. As the psychological research shows us, our decision making and creativity benefits from hearing outside perspectives and drawing parallels between different areas of interest; if we mine the same vein day after day, we may begin to pay less attention to the nuances and details. The German language, incidentally, has a word for this: the *Fachidiot*, a one-track specialist who takes a single-minded, inflexible approach to a multifaceted problem.

But perhaps the most pervasive—and potent—source of functional stupidity is the demand for complete corporate loyalty and an excessive focus on positivity, where the very idea of criticism may be seen as a betrayal, and admitting disappointment or anxiety is considered a weakness. This is a particular bugbear for Spicer, who told me that relentless optimism is now deeply embedded in many business cultures, stretching from start-ups to huge multinationals.

He described research on entrepreneurs, for instance, who often cling to the motto that they will "fail forward" or "fail early, fail often." Although these mottos sound like an example of the "growth mind-set"—which should improve your chances of success in the future—Spicer says that entrepreneurs often look to explain their failings with external factors ("my idea was before its time") rather than considering the errors in their own performance, and how it might be adapted in the future. They aren't really considering their own personal growth.

The numbers are huge: between 75 and 90 percent of entrepreneurs lose their first businesses—but by striving to remain relentlessly upbeat and positive, they remain oblivious to their mistakes.[9] "Instead of getting better—which this 'fail forward' idea would suggest—they actually get worse over time," Spicer said. "Because of these self-serving biases, they just go and start a new venture and make exactly the same mistakes over and over again . . . and they actually see this as a virtue."

The same attitude is prevalent among much larger and more established corporations, where bosses tell their employees to "only bring me the good news." Or you may attend a brainstorming session, where you are told that "no idea is a bad idea." Spicer argues that this is counterproductive; we are actually more creative when we take on board a criticism at an early stage of a discussion. "You've tested the assumptions and then you are able to enact upon them, instead of trying to push together ideas to cover up any differences."

~

I hope you will now understand the intelligence trap well enough to see immediately some of the dangers of this myopic approach.

The lack of curiosity and insight is particularly damaging during times of uncertainty. Based on his observations in editorial meetings, for instance, Alvesson has argued that overly rigid and unquestioning thinking of this kind prevented newspapers from exploring how factors like the economic climate and rising taxes were influencing their sales; editors were so fixated on examining specific headlines on their front pages that they forgot even to consider the need to explore broader new strategies or outlets for their stories.

But Nokia's implosion in the early 2010s offers the most vivid illustration of the ways that functional stupidity can drive an outwardly successful organization to failure.

If you owned a cellphone in the early 2000s, chances are that it was made by the Finnish company. In 2007, they held around half the global market share. Six years later, however, most of their customers had turned away from the clunky Nokia interface to more sophisticated smartphones, notably Apple's iPhone.

Commentators at the time suggested that Nokia was simply an inferior company with less talent and innovation than Apple, that the corporation had been unable to see the iPhone coming, or that they had been complacent, assuming that their own products would trump any others.

But as they investigated the company's demise, the Finnish and Singaporean researchers Timo Vuori and Quy Huy found that none of this was true.[10] Nokia's engineers were among the best in the world, and they were fully aware of the risks ahead. Even the CEO himself had admitted, during an interview, that he was "paranoid about all the competition." Yet they nevertheless failed to rise to the occasion.

One of the biggest challenges was Nokia's operating system, Symbian, which was inferior to Apple's iOS and unsuitable for dealing with sophisticated touchscreen apps, but overhauling the existing software would take years of development, and the management wanted to be able to present their new products quickly, leading them to rush through projects that needed greater forward planning.

Unfortunately, employees were not allowed to express any doubts about the way the company was proceeding. Senior managers would regularly shout "at the tops of their lungs" if you told them something they did not want to hear. Raise a doubt, and you risked losing your job. "If you were too negative, it would be your head on the block," one middle manager told the researchers. "The mind-set was that if you criticize what's being done, then you're not genuinely committed to it," said another.

As a consequence, employees began to feign expertise rather than admitting their ignorance about the problems they were facing, and accepted deadlines that they knew would be impossible to maintain.

They would even massage the data showing their results so as to give a better impression. And when the company lost employees, it deliberately hired replacements with a "can do" attitude—people who would nod along with new demands rather than disagreeing with the status quo. The company even ignored advice from external consultants, one of whom claimed that "Nokia has always been the most arrogant company ever towards my colleagues." They lost any chance of an outside perspective.

The very measures that were designed to focus the employee's attention and encourage a more creative outlook were making it harder and harder for Nokia to step up to the competition.

As a result, the company consistently failed to upgrade its operating system to a suitable standard—and the quality of Nokia's products slowly deteriorated. By the time the company launched the N8—their final attempt at an "iPhone Killer"—in 2010, most employees had secretly lost faith. It flopped, and after further losses Nokia's mobile phone business was acquired by Microsoft in 2013.

~

The concept of functional stupidity is inspired by extensive observational studies, including an analysis of Nokia's downfall, rather than psychological experiments, but this kind of corporate behavior shows clear parallels with psychologists' work on dysrationalia, wise reasoning and critical thinking.

You might remember, for instance, that feelings of threat trigger the so-called "hot," self-serving cognition that leads us to justify our own positions rather than seeking evidence that challenges our point of view—and this reduces scores of wise reasoning. (It is the reason we are wiser when advising a friend about a relationship problem, even if we struggle to see the solution to our own troubles.)

Led by its unyielding top management, Nokia as an organization was therefore beginning to act like an individual, faced with uncertain circumstances, whose ego has been threatened. Nokia's previous successes, meanwhile, may have given it a sense of "earned dogmatism," meaning that managers were less open to suggestions from experts outside the company.

Various experiments from social psychology suggest that this is a common pattern: groups under threat tend to become more conformist, single-minded and inward looking. More and more members begin to adopt the same views, and they start to favor simple messages over complex, nuanced ideas. This is even evident at the level of entire nations: newspaper editorials within a country tend to become more simplified and repetitive when it faces international conflict, for instance.[11]

No organization can control its external environment: some threats will be inevitable. But organizations can alter the way they translate those perceived dangers to employees, by encouraging alternative points of view and actively seeking disconfirming information. It's not enough to assume that employing the smartest people possible will automatically translate to better performance; you need to create the environment that allows them to use their skills.

Even the companies that appear to buck these trends may still incorporate some elements of evidence-based wisdom—although it may not be immediately obvious from their external reputation. The media company Netflix, for instance, famously has the motto that "adequate performance earns a generous severance"—a seemingly cut-throat attitude that might promote myopia and short-term gains over long-term resilience.

Yet they seem to balance this with other measures that are in line with the broader psychological research. A widely circulated presentation outlining Netflix's corporate vision, for example, emphasises many of the elements of good reasoning that we have discussed so far, including the need to recognize ambiguity and uncertainty and to challenge prevailing opinions—exactly the kind of culture that should encourage wise decision making.[12]

We can't, of course, know how Netflix will fare in the future. But its success to date would suggest that you can avoid functional stupidity while also running an efficient—some would say ruthless—operation.

The dangers of functional stupidity do not end with these instances of corporate failure. Besides impairing creativity and problem solving, a failure to encourage reflection and internal feedback can also lead to human tragedy, as NASA's disasters show.

"Often it leads to a number of small mistakes being made, or the [company] focuses on the wrong problems and overlooks a problem where there should have been some sort of post mortem," notes Spicer. As a consequence, an organization may appear outwardly successful while slowly sliding towards disaster.

Consider the Space Shuttle *Columbia* disaster in 2003, when foam insulation broke off an external tank during launch and struck the left wing of the orbiter. The resulting hole caused the shuttle to disintegrate upon re-entry into the Earth's atmosphere, leading to the death of all seven crew members.

The disaster would have been tragic enough had it been a fluke, a one-off occurrence without any potential warning signs. But NASA engineers had long known the insulation could break away like this; it had happened in every previous launch. For various reasons, however, the damage had never occurred in the right place to cause a crash, meaning that the NASA staff began to ignore the danger it posed.

"It went from being a troublesome event for engineers and managers to being classified as a housekeeping matter," Catherine Tinsley, a professor of management at Georgetown University in Washington DC who has specialized in studying corporate catastrophes, told me.

Amazingly, similar processes were also the cause of the *Challenger* crash in 1986, which exploded due to a faulty seal that had deteriorated in the cold Florida winter. Subsequent reports showed that the seals had cracked on many previous missions, but rather than see this as a warning, the staff had come to assume that it would always be safe. As Richard Feynman—a member of the Presidential Commission investigating the disaster—noted, "when playing Russian roulette, the fact that the first shot got off safely is little comfort for the next."[13] Yet NASA did not seem to have learned from those lessons.

Tinsley emphasises that this isn't a criticism of those particular engineers and managers. "These are really smart people, working with data, and trying really hard to do a good job." But NASA's errors demonstrate just how easily your perception of risk radically shifts without your even recognizing that a change has occurred. The organization was blind to the possibility of disaster.

The reason appears to be a form of cognitive miserliness known as the outcome bias, which leads us to focus on the actual consequences of a decision without even considering the alternative possible results. Like many of the other cognitive flaws that afflict otherwise intelligent people, it's really a lack of imagination: we passively accept the most salient detail from an event (what actually happened) and don't stop to think about what might have been, had the initial circumstances been slightly different.

Tinsley has now performed many experiments confirming that the outcome bias is a very common tendency among many different professionals. One study asked business students, NASA employees, and space-industry contractors to evaluate the mission controller "Chris," who took charge of an unmanned spacecraft under three different scenarios. In the first, the spacecraft launches perfectly, just as planned. In the second, it has a serious design flaw, but thanks to a turn of luck (its alignment to the sun) it can make its readings effectively. And in the third, there is no such stroke of fortune, and it completely fails.

Unsurprisingly, the complete failure is judged most harshly, but most of the participants were happy to ignore the design flaw in the "near-miss" scenario, and instead praised Chris's leadership skills. Importantly—and in line with Tinsley's theory that the outcome bias can explain disasters like the *Columbia* catastrophe—the perception of future dangers also diminished after the participants had read about the near miss, explaining how some organizations may slowly become immune to failure.[14]

Tinsley has now found that this tendency to overlook errors was the common factor in dozens of other catastrophes. "Multiple near-misses preceded and foreshadowed every disaster and business crisis we studied," Tinsley's team concluded in an article for the *Harvard Business Review* in 2011.[15]

Take one of the car manufacturer Toyota's biggest disasters. In August 2009, a Californian family of four died when the accelerator pedal of their Lexus jammed, leading the driver to lose control on the motorway and plough into an embankment at 120 miles per hour, where the car burst into flames. Toyota had to recall more than six million cars—a disaster that could have been avoided if the company had paid serious attention to more than two thousand reports of accelerator malfunction over the previous decades, which is around five times the number of complaints that a car manufacturer might normally expect to receive for this issue.[16]

Tellingly, Toyota had set up a high-level task force in 2005 to deal with quality control, but the company disbanded the group in early 2009, claiming that quality "was part of the company's DNA and therefore they didn't need a special committee to enforce it." Senior management also turned a deaf ear to specific warnings from more junior executives, while focusing on rapid corporate growth.[17] This was apparently a symptom of a generally insular way of operating that did not welcome outside input, in which important decisions were made only by those at the very top of the hierarchy. Like Nokia's management, it seems they simply didn't want to hear bad news that might sidetrack them from their broader goals.

The ultimate cost to Toyota's brand was greater than any of the savings they imagined they would make by not heeding these warnings. By 2010, 31 percent of Americans believed that Toyota cars were unsafe[18]—a dramatic fall from grace for a company that was once renowned for its products' quality and customer satisfaction.

Or consider Air France Flight 4590 from Paris to New York City. As it prepared for takeoff on July 25, 2000, the *Concorde* airliner ran over some sharp debris left on the runway, causing a 4.5 kg chunk of tyre to fly into the underside of the aircraft's wing. The resulting shockwave ruptured a fuel tank, leading it to catch fire during takeoff. The plane crashed into a nearby hotel, killing 113 people in total. Subsequent analyses revealed 57 previous instances in which the *Concorde* tire had burst on the runway, and in one case the damage was very nearly the same as for Flight 4590—except, through sheer good luck, the leaking fuel had failed to ignite. Yet these near misses were not taken as serious warning signs requiring urgent action.[19]

These crises are dramatic case studies in high-risk industries, but Tinsley argues that the same thinking processes will present latent dangers for many other organizations. She points to research on workplace safety, for instance, showing that for every thousand near misses, there will be one serious injury or fatality and at least ten smaller injuries.[20]

Tinsley does not frame her work as an example of "functional stupidity," but the outcome bias appears to arise from the same lack of reflection and curiosity that Spicer and Alvesson have outlined.

And even small changes to a company's environment can increase the chances that near misses are spotted. In both lab experiments and data gathered during real NASA projects, Tinsley has found that people are far more likely to note and report near misses when safety is emphasised as part of the overall culture, in its mission statements—sometimes with as much as a five-fold increase in reporting.[21]

As an example, consider one of those scenarios involving the NASA manager planning the unmanned space mission. Participants told that "NASA, which pushes the frontiers of knowledge, must operate in a high-risk, risk-tolerant environment" were much less likely to notice the near miss. Those told that "NASA, as a highly visible organization, must operate in a high-safety, safety-first environment," in contrast, successfully identified the latent danger. The same was also true when the participants were told that they would need to justify their judgment to the board. "Then the near miss also looks more like the failure condition."

Remember we are talking about unconscious biases here: no participants had weighed it up and considered the near miss was worth ignoring; but unless they were prompted, they just didn't really think about it at all. Some companies may expect that the value of safety is already implicitly understood, but Tinsley's work demonstrates that it needs to be highly salient. It is telling that NASA's motto had been "Faster, Better, Cheaper" for most of the decade leading up to the *Columbia* disaster.

Before we end our conversation, Tinsley emphasises that some risks will be inevitable; the danger is when we are not even aware they exist. She recalls a seminar during which a NASA engineer raised his hand in frustration. "Do you not want us to take any risks?" he asked. "Space missions are inherently risky."

"And my response was that I'm not here to tell you what your risk tolerance should be. I'm here to say that when you experience a near miss, your risk tolerance will increase and you won't be aware of it." As the fate of the *Challenger* and *Columbia* missions shows, no organization can afford that blind spot.

~

In hindsight, it is all too easy to see how Deepwater Horizon became a hotbed of irrationality before the spill. By the time of the explosion, it was six weeks behind schedule, with the delay costing $1 million a day, and some staff were unhappy with the pressure they were subjected to. In one email, written six days before the launch, the engineer Brian Morel labeled it "a nightmare well that has everyone all over the place."

These are exactly the high-pressure conditions that are now known to reduce reflection and analytical thinking. The result was a collective blind spot that prevented many of Deepwater Horizon's employees (from BP and its partners, Halliburton and Transocean) from seeing the disaster looming, and contributed to a series of striking errors.

To try to reduce the accumulating costs, for instance, they chose to use a cheaper mix of cement to secure the well, without investigating the possibility that it may not have been stable enough for the job at hand. They also reduced the total volume of cement used—violating their own guidelines—and scrimped on the necessary equipment required to hold the well in place.

On the day of the accident itself, the team avoided completing the full suite of tests to ensure the seal was secure, while also ignoring anomalous results that might have predicted the build-up of pressure inside the well.[22] Worse still, the equipment necessary to contain the blowout, once it occurred, was in ill-repair.

Each of these risk factors could have been identified long before disaster struck; as we have seen, there were many minor blowouts that should have been significant warnings of the underlying dangers, leading to new and updated safety procedures. Thanks to lucky circumstances, however—even the random direction of the wind—none had been fatal, and so the underlying factors, including severe corner-cutting and inadequate safety training, had not been examined.[23] And the more they played with fate, the more they were lulled into a false sense of complacency and became less concerned about cutting corners.[24] It was a classic case of the outcome bias that Tinsley has documented—and the error seemed to have been prevalent across the whole of the oil industry.

Eight months previously, another oil and gas company, PTT, had even witnessed a blowout and spill in the Timor Sea, off Australia. Halliburton, which had also worked on the Macondo well, was the company behind the cement job there, too, and although a subsequent report had claimed that Halliburton itself held little responsibility, it might have still been taken as a vivid reminder of the dangers involved. A lack of communication between operators and experts, however, meant the lessons were largely ignored by the Deepwater Horizon team.[25]

In this way, we can see that the disaster wasn't down to the behavior of any one employee, but to an endemic lack of reflection, engagement, and critical thinking that meant decision makers across the project had failed to consider the true consequences of their actions.

"It is the underlying 'unconscious mind' that governs the actions of an organization and its personnel," a report from the Center for Catastrophic Risk Management (CCRM) at the University of California, Berkeley, concluded.[26] "These failures . . . appear to be deeply rooted in a multi-decade history of organizational malfunction and short-sightedness." In particular, the management had become so obsessed with pursuing further success, they had forgotten their own fallibilities and the vulnerabilities of the technology they were using. They had "forgotten to be afraid."

Or as Karlene Roberts, the director of the CCRM, told me in an interview, "Often, when organizations look for the errors that caused something catastrophic to happen, they look for someone to name, blame, and then train or get rid of. . . . But it's rarely what happened on the spot that caused the accident. It's often what happened years before."

If this "unconscious mind" represents an organizational intelligence trap, how can an institution wake up to latent risks?

In addition to studying disasters, Roberts's team has also examined the common structures and behaviors of "high-reliability organizations" such as nuclear power plants, aircraft carriers, and air traffic control systems that operate with enormous uncertainty and potential for hazard, yet somehow achieve extremely low failure rates.

Much like the theories of functional stupidity, their findings emphasise the need for reflection, questioning, and the consideration of long-term consequences—including, for example, policies that give employees the "licence to think."

Refining these findings to a set of core characteristics, Karl Weick and Kathleen Sutcliffe have shown that high-reliability organizations all demonstrate:[27]

- **Preoccupation with failure:** The organization does not become complacent with success, and workers assume "each day will be a bad day." The organization rewards employees for self-reporting errors.

- **Reluctance to simplify interpretations:** Employees are rewarded for questioning assumptions and for being skeptical of received wisdom. At Deepwater Horizon, for instance, more engineers and managers may have raised concerns about the poor quality of the cement and asked for further tests.

- **Sensitivity to operations:** Team members continue to communicate and interact, to update their understanding of the situation at hand and search for the root causes of anomalies. On Deepwater Horizon, the rig staff should have been more curious about the anomalous pressure tests, rather than accepting the first explanation.

- **Commitment to resilience:** Building the necessary knowledge and resources to bounce back after error occurs, including regular "pre-mortems" and regular discussions of near misses. Long before the Deepwater Horizon explosion, BP might have examined the underlying organizational factors leading to previous, less serious accidents, and ensured all team members were adequately prepared to deal with a blowout.

- **Deference to expertise:** This relates to the importance of communication between ranks of the hierarchy, and the intellectual humility of those at the top. Executives need to trust the people on the ground. Toyota and NASA, for instance, both failed to heed the concerns of engineers; similarly, after the Deepwater Horizon explosion, the media reported that workers at BP had been scared of raising concerns in case they would be fired.[28]

The commitment to resilience may be evident in small gestures that allow workers to know that their commitment to safety is valued. On one aircraft carrier, the USS *Carl Vinson*, a crewmember reported that he had lost a tool on deck that could have been sucked into a jet engine. All aircraft were redirected to land—at significant cost—but rather than punishing the team member for his carelessness, he was commended for his honesty in a formal ceremony the next day. The message was clear—errors would be tolerated if they were reported, meaning that the team as a whole were less likely to overlook much smaller mistakes.

The US Navy, meanwhile, has employed the SUBSAFE system to reduce accidents on its nuclear submarines. The system was first implemented following the loss of the USS *Thresher* in 1963, which flooded due to a poor joint in its pumping system, resulting in the deaths of 112 Navy personnel and 17 civilians.[29] SUBSAFE specifically instructs officers to experience "chronic uneasiness," summarized in the saying "trust, but verify," and in more than five decades since, they haven't lost a single submarine using the system.[30]

Inspired by Ellen Langer's work, Weick refers to these combined characteristics as "collective mindfulness." The underlying principle is that the organization should implement any measures that encourage its employees to remain attentive, proactive, open to new ideas, questioning of every possibility, and devoted to discovering and learning from mistakes, rather than simply repeating the same behaviors over and over.

There is good evidence that adopting this framework can result in dramatic improvements. Some of the most notable successes of applying collective mindfulness have come from healthcare. (We've already seen how doctors are changing how individuals think—but this specifically concerns the overall culture and group reasoning.) The available measures involve empowering junior staff to question assumptions and to be more critical of the evidence presented to them, and encouraging senior staff to actively engage the opinions of those beneath them so that everyone is accountable to everyone else. The staff also have regular "safety huddles," proactively report errors and perform detailed "root-cause analyses" to examine the underlying processes that may have contributed to any mistake or near miss.

Using such techniques, one Canadian hospital, St. Joseph's Healthcare in London, Ontario, has reduced medication errors (the wrong drugs given to the wrong person) to just two mistakes in more than 800,000 medications dispensed in the second quarter of 2016. The Golden Valley Memorial in Missouri, meanwhile, has reduced drug-resistant *Staphylococcus aureus* infections to zero using the same principles, and patient falls—a serious cause of unnecessary injury in hospitals—have dropped by 41 percent.[31]

Despite the additional responsibilities, staff in mindful organizations often thrive on the extra workload, with a lower turnover rate than institutions that do not impose these measures.[32] Contrary to expectations, it is more rewarding to feel like you are fully engaging your mind for the greater good, rather than simply going through the motions.

~

In these ways, the research on functional stupidity and mindful organizations perfectly complement each other, revealing the ways that our environment can either involve the group brain in reflection and deep thinking, or dangerously narrow its focus so that it loses the benefits of its combined intelligence and expertise. They offer us a framework to understand the intelligence trap and evidence-based wisdom on a grand scale.

Beyond these general principles, the research also reveals specific practical steps for any organization hoping to reduce error. Given that our biases are often amplified by feelings of time pressure, Tinsley suggests that organizations should encourage employees to examine their actions and ask: "If I had more time and resources, would I make the same decisions?" She also believes that people working on high-stakes projects should take regular breaks to "pause and learn," where they may specifically look for near misses and examine the factors underlying them—a strategy, she says, that NASA has now applied. They should institute near-miss reporting systems; "and if you don't report a near miss, you are then held accountable."

Spicer, meanwhile, proposes adding regular reflective routines to team meetings, including pre-mortems and post-mortems, and appointing a devil's advocate whose role is to question decisions and look for flaws in their logic. "There's lots of social psychology that says it leads to slightly dissatisfied people but better-quality decisions." He also recommends taking advantage of the outside perspective, by either inviting secondments from other companies, or encouraging staff to shadow employees from other organizations and other industries, a strategy that can help puncture the bias blind spot.

The aim is to do whatever you can to embrace that "chronic uneasiness"—the sense that there might always be a better way of doing things.

Looking to research from further afield, organizations may also benefit from tests such as Keith Stanovich's rationality quotient, which would allow them to screen employees working on high-risk projects and to check whether they are more or less susceptible to bias, and if they are in need of further training. They might also think of establishing critical thinking programs within the company.

They may also analyze the mind-set embedded in its culture:

whether it encourages the growth of talent or leads employees to believe that their abilities are set in stone. Carol Dweck's team of researchers asked employees at seven *Fortune* 1000 companies to rate their level of agreement with a series of statements, such as: "When it comes to being successful, this company seems to believe that people have a certain amount of talent, and they really can't do much to change it" (reflecting a collective fixed mind-set) or "This company genuinely values the development and growth of its employees" (reflecting a collective growth mind-set).

As you might hope, companies cultivating a collective growth mind-set enjoyed greater innovation and productivity, more collaboration within teams and higher employee commitment. Importantly, employees were also less likely to cut corners, or cheat to get ahead. They knew their development would be encouraged and were therefore less likely to cover up for their perceived failings.[33]

During their corporate training, organizations could also make use of productive struggle and desirable difficulties to ensure that their employees process the information more deeply. As we saw in Chapter 8, this not only means that the material is recalled more readily; it also increases overall engagement with the underlying concepts and means that the lessons are more readily transferable to new situations.

Ultimately, the secrets of wise decision making for the organization are very similar to the secrets of wise decision making for the intelligent individual. Whether you are a forensic scientist, doctor, student, teacher, financier, or aeronautical engineer, it pays to humbly recognize your limits and the possibility of failure, take account of ambiguity and uncertainty, remain curious and open to new information, recognize the potential to grow from errors, and actively question everything.

~

In the Presidential Commission's damning report on the Deepwater Horizon explosion, one particular recommendation catches the attention, inspired by a revolutionary change in US nuclear power plants as a model for how an industry may deal with risk more mindfully.[34]

As you might have come to expect, the trigger was a real crisis. ("Everyone waits to be punished before they act," Roberts said.) In this case it was the partial meltdown of a radioactive core in the Three Mile Island Nuclear Generating Station in 1979. The disaster led to the foundation of a new regulator, the Institute of Nuclear Power Operations (INPO), which incorporates a number of important characteristics.

Each generator is visited by a team of inspectors every two years, each visit lasting five to six weeks. Although one-third of INPO's inspectors are permanent staff, the majority are seconded from other power plants, leading to a greater sharing of knowledge between organizations, and the regular input of an outside perspective in each company. INPO also actively facilitates discussions between lower-level employees and senior management with regular review groups. This ensures that the fine details and challenges of day-to-day operations are acknowledged and understood at every level of the hierarchy.

To increase accountability, the results of the inspections are announced at an annual dinner—meaning that "You get the whole top level of the utility industry focused on the poor performer," according to one CEO quoted in the Presidential Commission's report. Often, CEOs in the room will offer to loan their expertise to bring other generators up to scratch. The result is that every company is constantly learning from each other's mistakes. Since INPO began operating, US generators have seen a tenfold reduction in the number of worker accidents.[35]

You need not be a fan of nuclear power to see how these structures maximize the collective intelligence of employees across the industry and greatly increase each individual's awareness of potential risks, while reducing the build-up of those small, unacknowledged errors that can lead to catastrophe. INPO shows the way that regulatory bodies can help mindful cultures to spread across organizations, uniting thousands of employees in their reflection and critical thinking.

The oil industry has not (yet) implemented a comparably intricate system, but energy companies have banded together to revise industry standards, improve worker training and education, and upgrade their technology to better contain a spill, should it occur. BP has also funded a huge research program to deal with the environmental devastation in the Gulf of Mexico. Some lessons have been learned—but at what cost?[36]

~

The intelligence trap often emerges from an inability to think beyond our expectations—to imagine an alternative vision of the world, where our decision is wrong rather than right. This must have been the case on April 20, 2010; no one can possibly have considered the true scale of the catastrophe they were letting loose.

Over the subsequent months, the oil slick would cover more than 43,000 square miles of the ocean's surface—an area that is roughly 85 percent the size of England.[37] According to the Center for Biological Diversity, the disaster killed at least 80,000 birds, 6,000 sea turtles and 26,000 marine mammals—an ecosystem destroyed by preventable errors. Five years later, baby dolphins were still being born with underdeveloped lungs, due to the toxic effects of the oil leaked into the water and the poor health of their parents. Only 20 percent of dolphin pregnancies resulted in a live birth.[38]

That's not to mention the enormous human cost. Besides the eleven lives lost on the rig itself and the unimaginable trauma inflicted on those who escaped, the spill devastated the livelihoods of fishing communities in the Gulf. Two years after the spill, Darla Rooks, a lifelong fisherperson from Port Sulfur, Louisiana, described finding crabs "with holes in their shells, shells with all the points burned off so all the spikes on their shells and claws are gone, misshapen shells, and crabs that are dying from within . . . they are still alive, but you open them up and they smell like they've been dead for a week."

The level of depression in the area rose by 25 percent over the following months, and many communities struggled to recover from their losses. "Think about losing everything that makes you happy, because that is exactly what happens when someone spills oil and sprays dispersants on it," Rooks told Al Jazeera in 2012.[39] "People who live here know better than to swim in or eat what comes out of our waters."

This disaster was entirely preventable—if only BP and its partners had recognized the fallibility of the human brain and its capacity for error. No one is immune, and the dark stain in the Gulf of Mexico should be a constant reminder of the truly catastrophic potential of the intelligence trap.

Epilogue

We began this journey with the story of Kary Mullis—the brilliant chemist who has dabbled in astrology and astral projection, and even defended AIDS denialism. It should now be clear how factors such as motivated reasoning could have led him to ignore every warning sign.

But I hope it has become clear that *The Intelligence Trap* is so much more than the story of any individual's mistakes. The trap is a phenomenon that concerns us all, given the kinds of thinking that we, as a society, have come to appreciate, and the ones we have neglected.

Interviewing so many brilliant scientists for this book, I came to notice that each expert seemed, in some way, to embody the kind of intelligence or thinking that they've been studying. David Perkins was unusually thoughtful, frequently pausing our conversation to reflect before we continued; Robert Sternberg, meanwhile, was tremendously pragmatic in conveying his message; Igor Grossmann was extremely humble and took extra care to emphasise the limits of his knowledge; and Susan Engel was animated with endless curiosity.

Perhaps they were attracted to their field because they wanted to understand their own thinking better; or perhaps their own thinking came to resemble the subject of their study. Either way, to me it was one more illustration of the enormous range of potential thinking styles available to us, and the benefits they bring.

James Flynn describes the rise in IQ over the twentieth century as our "cognitive history"; it shows the ways our minds have been molded by the society around us. But it strikes me that if each of these scientists had been able to present and promote their work in the early nineteenth century, before the concept of general intelligence came to determine the kind of thinking that was considered "smart," our cognitive history might have been very different. As it is, the abstract reasoning measured by IQ tests, SATs, and GREs still dominates our understanding of what constitutes intelligence.

We don't need to deny the value of those skills, or abandon the learning of factual knowledge and expertise, to accept that other ways of reasoning and learning are equally deserving of our attention. Indeed, if I have learned anything from this research, it is that cultivating these other traits often enhances the skills measured by standard tests of cognitive ability, as well as making us more rounded and wiser thinkers.

Study after study has shown that encouraging people to define their own problems, explore different perspectives, imagine alternative outcomes to events, and identify erroneous arguments can boost their overall capacity to learn new material while also encouraging a wiser way of reasoning.[1]

I found it particularly encouraging that learning with these methods often benefits people across the intelligence spectrum. They can reduce motivated reasoning among the highly intelligent, for instance, but they can also improve the general learning of people with lower intelligence. One study by Bradley Owens at the State University of New York in Buffalo, for instance, found that intellectual humility predicted academic achievement better than an IQ test. Everyone with higher intellectual humility performed better, but—crucially—it was of most benefit for those with lower intelligence, completely compensating for their lower "natural" ability.[2] The principles of evidence-based wisdom can help anyone to maximise their potential.

~

This new understanding of human thinking and reasoning could not have come at a more important time.

As Robert Sternberg wrote in 2018: "The steep rise in IQ has bought us, as a society, much less than anyone had any right to hope for. People are probably better at figuring out complex cell phones and other technological innovations than they would have been at the turn of the twentieth century. But in terms of our behavior as a society, are you impressed with what 30 points has brought us?"[3]

Although we have made some strides in areas such as technology and healthcare, we are no closer to solving pressing issues such as climate change or social inequality—and the increasingly dogmatic views that often come with the intelligence trap only stand in the way of the negotiations between people with different positions that might lead to a solution. The World Economic Forum has listed increasing political polarization and the spread of misinformation in "digital wildfires"[4] as two of the greatest threats facing us today—comparable to terrorism and cyber warfare.

The twenty-first century presents complex problems that require a wiser way of reasoning, one that recognizes our current limitations, tolerates ambiguity and uncertainty, balances multiple perspectives, and bridges diverse areas of expertise. And it is becoming increasingly clear that we need more people who embody those qualities.

This may sound like wishful thinking, but remember that American presidents who scored higher on scales of open-mindedness and perspective taking were far more likely to find peaceful solutions to conflict. It's not unreasonable to ask whether, given this research, we should be actively demanding those qualities in our leaders, in addition to more obvious measures of academic achievement and professional success.

~

If you want to apply this research yourself, the first step is to acknowledge the problem. We have now seen how intellectual humility can help us see through our bias blind spot, form more rational opinions, avoid misinformation, learn more effectively, and work more productively with the people around us. As the philosopher Valerie Tiberius, who is now working with psychologists at the Chicago Center for Practical Wisdom, points out, we often spend huge amounts of time trying to boost our self-esteem and confidence. "But I think that if more people had some humility about what they know and don't know, that would go a tremendous distance to improving life for everyone."

To this end, I have included a short "taxonomy" of definitions in the appendix, outlining the most common errors at the heart of the intelligence trap and some of the best ways to deal with them. Sometimes, just being able to put a label on your thinking opens the door to a more insightful frame of mind. I have found that it can be an exhilarating experience to question your own intelligence in these ways, as you reject many of the assumptions you have always taken for granted. It allows you to revive the childlike joy of discovery that drove everyone from Benjamin Franklin to Richard Feynman.

It is easy, as adults, to assume that we have reached our intellectual peak by the time we finish our education; indeed, we are often told to expect a mental decline soon after. But the work on evidence-based wisdom shows that we can all learn new ways of thinking. Whatever our age and expertise, whether a NASA scientist or a school student, we can all benefit from wielding our minds with insight, precision, and humility.[5]

Appendix: Taxonomies of Stupidity and Wisdom

A Taxonomy of Stupidity

Bias blind spot: Our tendency to see others' flaws, while being oblivious to the prejudices and errors in our own reasoning.

Cognitive miserliness: A tendency to base our decision making on intuition rather than analysis.

Contaminated mindware: An erroneous baseline knowledge that may then lead to further irrational behavior. Someone who has been brought up to distrust scientific evidence may then be more susceptible to quack medicines and beliefs in the paranormal, for instance.

Dysrationalia: The mismatch between intelligence and rationality, as seen in the life story of Arthur Conan Doyle. This may be caused by cognitive miserliness or contaminated mindware.

Earned dogmatism: Our self-perceptions of expertise mean we have gained the right to be closed-minded and to ignore other points of view.

Entrenchment: The process by which an expert's ideas become rigid and fixed.

Fachidiot: Professional idiot. A German term to describe a one-track specialist who is an expert in their field but takes a blinkered approach to a multifaceted problem.

Fixed mind-set: The belief that intelligence and talent are innate, and exerting effort is a sign of weakness. Besides limiting our ability to learn, this attitude also seems to make us generally more closed-minded and intellectually arrogant.

Functional stupidity: A general reluctance to self-reflect, question our assumptions, and reason about the consequences of our actions. Although this may increase productivity in the short term (making it "functional"), it reduces creativity and critical thinking in the long term.

"Hot" cognition: Reactive, emotionally charged thinking that may give full rein to our biases. Potentially one source of Solomon's paradox (see below).

Meta-forgetfulness: A form of intellectual arrogance. We fail to keep track of how much we know and how much we have forgotten; we assume that our current knowledge is the same as our peak knowledge. This is common among university graduates; years down the line, they believe that they understand the issues as well as they did when they took their final exams.

Mindlessness: A lack of attention and insight into our actions and the world around us. It is a particular issue in the way children are educated.

Moses illusion: A failure to spot contradictions in a text, due to its fluency and familiarity. For example, when answering the question, "How many animals of each kind did Moses take on the Ark?," most people answer two. This kind of distraction is a common tactic for purveyors of misinformation and fake news.

Motivated reasoning: The unconscious tendency to apply our brainpower only when the conclusions will suit our predetermined goal. It may include the confirmation or myside bias (preferentially seeking and remembering information that suits our goal) and discomfirmation bias (the tendency to be especially skeptical about evidence that does not fit our goal). In politics, for instance, we are far more likely to critique evidence concerning an issue such as climate change if it does not fit with our existing worldview.

Peter principle: We are promoted based on our aptitude at our current job—not on our potential to fill the next role. This means that managers inevitably "rise to their level of incompetence." Lacking the practical intelligence necessary to manage teams, they subsequently underperform. (Named after management theorist Laurence Peter.)

Pseudo-profound bullshit: Seemingly impressive assertions that are presented as true and meaningful but are actually vacuous under further consideration. Like the Moses illusion, we may accept their message due to a general lack of reflection.

Solomon's paradox: Named after the ancient Israelite king, Solomon's paradox describes our inability to reason wisely about our own lives, even if we demonstrate good judgment when faced with other people's problems.

Strategic ignorance: Deliberately avoiding the chance to learn new information to avoid discomfort and to increase our productivity. At work, for instance, it can be beneficial not to question the long-term consequences of your actions, if that knowledge will interfere with the chances of promotion. These choices may be unconscious.

The too-much-talent effect: The unexpected failure of teams once their proportion of "star" players reaches a certain threshold. See, for instance, the England football team in the Euro 2016 tournament.

A Taxonomy of Wisdom

Actively open-minded thinking: The deliberate pursuit of alternative viewpoints and evidence that may question our opinions.

Cognitive inoculation: A strategy to reduce biased reasoning by deliberately exposing ourselves to examples of flawed arguments.

Collective intelligence: A team's ability to reason as one unit. Although it is *very* loosely connected to IQ, factors such as the social sensitivity of the team's members seem to be far more important.

Desirable difficulties: A powerful concept in education: we actually learn better if our initial understanding is made harder, not easier. See also Growth mind-set.

Emotional compass: A combination of interoception (sensitivity to bodily signals), emotion differentiation (the capacity to label your feelings in precise detail), and emotion regulation that together help us to avoid cognitive and affective biases.

Epistemic accuracy: Someone is epistemically accurate if their beliefs are supported by reason and factual evidence.

Epistemic curiosity: An inquisitive, interested, questioning attitude; a hunger for information. Not only does curiosity improve learning; the latest research shows that it also protects us from motivated reasoning and bias.

Foreign language effect: The surprising tendency to become more rational when speaking a second language.

Growth mind-set: The belief that talents can be developed and trained. Although the early scientific research on mind-set focused on its role in academic achievement, it is becoming increasingly clear that it may drive wiser decision making, by contributing to traits such as intellectual humility.

Intellectual humility: The capacity to accept the limits of our judgment and to try to compensate for our fallibility. Scientific research has revealed that this is a critical, but neglected, characteristic that determines much of our decision making and learning, and which may be particularly crucial for team leaders.

Mindfulness: The opposite of mindlessness. Although this can include meditative practice, it refers to a generally reflective and engaged state that avoids reactive, overly emotional responses to events and allows us to note and consider our intuitions more objectively. The term may also refer to an organization's risk management strategy (see Chapter 10).

Moral algebra: Benjamin Franklin's strategy to weigh up the pros and cons of an argument, often over several days. By taking this slow and systematic approach, you may avoid issues such as the availability bias—our tendency to base judgments on the first information that comes to mind—allowing you to come to a wiser long-term solution to your problem.

Pre-mortem: Deliberately considering the worst-case scenario, and all the factors that may have contributed towards it, before making a decision. This is one of the most well-established "de-biasing" strategies.

Reflective competence: The final stage of expertise, when we can pause and analyze our gut feelings, basing our decisions on both intuition and analysis. See also Mindfulness.

Socrates effect: A form of perspective taking, in which we imagine explaining our problem to a young child. The strategy appears to reduce "hot" cognition and reduce biases and motivated reasoning.

Tolerance of ambiguity: A tendency to embrace uncertainty and nuance, rather than seeking immediate closure on the issue at hand.

Solution to the Jack-Anne-George test (p. 4)

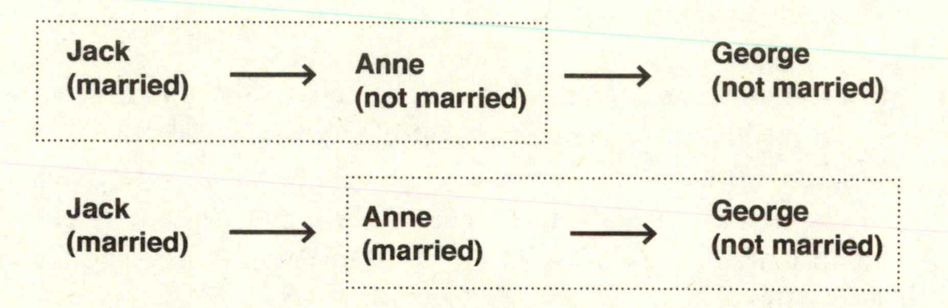

The trick is to think through both possible scenarios – whether Anne is married or not married. As you can see from the diagram above, in either case one married person will be looking at one unmarried person.

Notes

Introduction

1 These quotes can all be found in Mullis's autobiography: Mullis, K. (1999), *Dancing Naked in the Mind Field*, London: Bloomsbury. He also expresses them on his website: https://www.karymullis.com/pdf/On_AIDS_and_Global_Warming.pdf . They are frequently quoted on web pages and forums discussing astrology, climate change, and HIV/AIDS conspiracy theories. For instance, he is quoted on the website of the prominent AIDS denialist, Peter Duesberg: http://www.duesberg.com/viewpoints/kintro.html. And numerous video interviews of him describing AIDS conspiracy theories can be seen on YouTube: e.g. https://www.youtube.com/watch?v=IifgAvXU3ts&t=7s and https://www.youtube.com/watch?v=rycOLjoPbeo.

2 Graber, M.L. (2013), "The Incidence of Diagnostic Error in Medicine," *BMJ Quality and Safety*, 22(Suppl. 2), 21–7.

3 According to the classicist Christopher Rowe, these details of Socrates's appearance and life are broadly consistent across multiple sources. All of the quotations have also been taken from Rowe's translation of Plato's *Apology* contained in this volume: Rowe, C. (2010), *The Last Days of Socrates*, London: Penguin (Kindle Edition). The parallels between Socrates's trial and the research on the bias blind spot may not be the only instance in which Greek philosophy preempted behavioral economics and psychology. In an article for the online magazine *Aeon*, the journalist Nick Romeo finds examples of framing, confirmation bias, and anchoring in Plato's teaching. See Romeo, N. (2017), "Platonically Irrational," *Aeon*, https://aeon.co/essays/what-plato-knew-about-behavioural-economics-a-lot.

4 Descartes, R. (1637), *A Discourse on the Method*, trans. Maclean, I. (2006), Oxford: Oxford University Press, p. 5.

Chapter 1

1 Shurkin, J. (1992), *Terman's Kids: The Groundbreaking Study of How the Gifted Grow Up*, Boston, MA: Little, Brown, p. 122.

2 Shurkin, *Terman's Kids*, pp. 51–3.

3 Shurkin, *Terman's Kids*, pp. 109–16.

4 Shurkin, *Terman's Kids*, pp. 54–8.

5 Terman, L.M. (1922), "Were We Born That Way?" *World's Work*, 44, 657–9. Quoted in White, J. (2006), *Intelligence, Destiny and Education: The Ideological Roots of Intelligence Testing*, London: Routledge, p. 24.

6 Terman, L.M. (1930), "Trails to Psychology," in Murchison, C. (ed.), *History of Psychology in Autobiography*, Vol. 2, p. 297.

7 Terman, "Trails to Psychology," p. 303.

8 Nicolas, S., et al. (2013), "Sick or Slow? On the Origins of Intelligence as a Psychological Object," *Intelligence*, 41 (5), 699–711.

9 For more information on Binet's views, see White, S. H. (2000), "Conceptual Foundations of IQ Testing," *Psychology, Public Policy, and Law*, 6, 33–43.

10 Binet, A. (1909), *Les idées modernes sur les enfants*, Paris: Flammarion.

11 Perkins, D. (1995), *Outsmarting IQ: The Emerging Science of Learnable Intelligence*, New York: Free Press, p. 44.

12 Terman, L.M. (1916), *The Measurement of Intelligence: An Explanation of and a Complete Guide for the Use of the Stanford Revision and Extension of the Binet-Simon Intelligence Scale*, Boston, MA: Houghton Mifflin, p. 46.

13 Terman, *The Measurement of Intelligence*, p. 6.

14 Terman, *The Measurement of Intelligence*, p. 11.

15 Shurkin, *Terman's Kids*.

16 Shurkin, *Terman's Kids*, pp. 196–292.

17 Honan, W. (9 March 2002), "Shelley Mydans, 86, author and former POW," *New York Times*.

18 McGraw, C. (29 December 1988), "Creator of '"Lucy"' TV show dies," *Los Angeles Times*.

19 Oppenheimer, J., and Oppenheimer, G. (1996), *Laughs, Luck—and Lucy: How I Came to Create the Most Popular Sitcom of All Time*. Syracuse, NY: Syracuse University Press, p. 100.

20 Terman, "Trails to Psychology," p. 297.

21 In China, for instance, most schools maintain records of each child's performance on nonverbal reasoning tests. See Higgins, L.T. and Xiang, G. (2009), "The Development and Use of Intelligence Tests in China," *Psychology and Developing Societies*, 21(2), 257–75.

22 Madhok, D. (10 September 2012), "Cram Schools Boom Widens India Class Divide," Reuters, https://in.reuters.com/article/india-cramschools-kota/cram-schools-boom-widens-indias-class-divide-idINDEE8890GW20120910.

23 Ritchie, S.J., et al. (2015), "Beyond a Bigger Brain: Multivariable Structural Brain Imaging and Intelligence," *Intelligence*, 51, 47–56.

24 Gregory, M.D., Kippenhan, J.S., Dickinson, D., Carrasco, J., Mattay, V.S., Weinberger, D.R., and Berman, K.F. (2016), "Regional Variations in Brain Gyrification Are Associated with General Cognitive Ability in Humans," *Current Biology*, 26(10), 1301–5.

25 Li, Y., Liu, Y., Li, J., Qin, W., Li, K., Yu, C., and Jiang, T. (2009), "Brain Anatomical Network and Intelligence," *PLoS Computational Biology*, 5(5), e1000395.

26 For further discussion of this, see Kaufman, S. (2013), "*Ungifted: Intelligence Redefined*, New York: Basic Books (Kindle Edition). See, in particular, his review of the research by the Posse Foundation (pp. 286–8), who selected university students based on a wide variety of measures besides traditional, abstract intelligence—including in-depth interviews and group discussions that measure qualities such as leadership, communication, problem solving, and collaborative skills. Although their SAT scores are well below the norm for their university, their subsequent success at university is roughly equal to the average of other students.

27 The following paper, written by some of the most prominent IQ researchers, explicitly makes this point: Neisser, U., et al. (1996), "Intelligence: Knowns and Unknowns" *American Psychologist*, 51(2), 77–101. See also the following paper, which provides a further analysis of this idea, including the following statement: "Over one hundred years of research on intelligence testing has shown that scores on standardized tests of intelligence predict a wide range of outcomes, but even the strongest advocates of intelligence testing agree that IQ scores (and their near cousins such as the SATs) leave a large portion of the variance unexplained when predicting real-life behaviors." Butler, H.A., Pentoney, C., and Bong, M.P. (2017), "Predicting Real-World Outcomes: Critical Thinking Ability Is a Better Predictor of Life Decisions than Intelligence," *Thinking Skills and Creativity*, 25, 38–46.

28 Schmidt, F.L., and Hunter, J. (2004), "General Mental Ability in the World of Work: Occupational Attainment and Job Performance," *Journal of Personality and Social Psychology*, 86(1), 162–73.

29 Neisser, U., et al., "Intelligence." Strenze, T. (2007), "Intelligence and Socioeconomic Success: A Meta-Analytic Review of Longitudinal Research," *Intelligence*, 35, 401–26.

30 For a discussion of the difficulties with linking IQ to job performance, see: Byington, E., and Felps, W. (2010), "Why Do IQ Scores Predict Job Performance? An Alternative, Sociological Explanation," *Research in Organizational Behavior*, 30, 175–202. Richardson, K., and Norgate, S.H. (2015), "Does IQ Really Predict Job Performance?" *Applied Developmental Science*, 19(3), 153–69. Ericsson, K.A. (2014), "Why Expert Performance Is Special and Cannot Be Extrapolated from Studies of Performance in the General Population: A Response to Criticisms," *Intelligence*, 45, 81–103.

31 Feldman, D. (1984), "A Follow-Up of Subjects Scoring above 180 IQ in Terman's Genetic Studies of Genius," *Exceptional Children*, 50(6), 518–23.

32 Shurkin, *Terman's Kids*, pp. 183–7.

33 Shurkin, *Terman's Kids*, p. 190.

34 For a more recent analysis coming to broadly the same conclusions, see also Dean Simonton's analysis of the Terman studies of genius. "Not only do differences in general intelligence explain little variance in achieved eminence, but the explanatory power of intelligence is apparently also contingent on having intelligence defined in more domain specific terms. In essence, intellectual giftedness must be reconceived as the degree of acceleration in expertise acquisition within an individually chosen domain. Furthermore, personality differences and early developmental experiences have an even bigger part to play in the emergence of genius, although these influential factors must also be tailored to the specific domain of achievement." Simonton, D.K. (2016), "Reverse Engineering Genius: Historiometric Studies of Superlative Talent," *Annals of the New York Academy of Sciences*, 1377, 3–9.

35 Elements of this interview first appeared in an article I wrote for BBC Future in 2016: http://www.bbc.com/future/story/20160929-our-iqs-have-never-been-higher-but-it-hasnt-made-us-smart.

36 Clark, C.M., Lawlor-Savage, L., and Goghari, V.M. (2016), "The Flynn Effect: A Quantitative Commentary on Modernity and Human Intelligence," *Measurement: Interdisciplinary Research and Perspectives*, 14(2), 39–53. In line with the idea of scientific spectacles, recent research has shown that the Flynn Effect can largely be accounted for in the time people take to answer the questions. Younger generations do it more rapidly, as if abstract thinking has been automated and become second nature: Must, O. and Must, A. (2018), "Speed and the Flynn Effect," *Intelligence*, 68, 37–47.

37 Some modern IQ researchers have in fact suggested that training in these abstract thinking skills could be a way of closing the social divide between low- and high-IQ individuals. But the Flynn Effect would suggest that this would be of limited benefit for things such as creative thinking. See, for instance, Asbury, K., and Plomin, R. (2014), *G Is for Genes*, Oxford: Wiley Blackwell, pp. 149–87.

38 In fact, there is some evidence that creativity has actually decreased over the same period, both in terms of lab-based measures of creative problem solving and real-world measures of innovation, such as the average number of patents per person. See Kim, K.H. (2011), "The Creativity Crisis: The Decrease in Creative Thinking Scores on the Torrance Tests of Creative Thinking," *Creativity Research Journal*, 23(4), 285–95. Kaufman, J. (2018), "Creativity as a Stepping Stone toward a Brighter Future," *Journal of Intelligence*, 6(2), 21. Huebner, J. (2005), "A Possible Declining Trend for Worldwide Innovation," *Technological Forecasting and Social Change*, 72(8), 980–6.

39 Flynn, J.R. (1998), "IQ Gains over Time: Toward Finding the Causes," in Neisser, U. (ed.), *The Rising Curve: Long-Term Changes in IQ and Related Measures*, Washington, DC: American Psychological Association, pp. 25–66.

40 Harms, P.D., and Credé, M. (2010), "Remaining Issues in Emotional Intelligence Research: Construct Overlap, Method Artifacts, and Lack of Incremental Validity," *Industrial and Organizational Psychology: Perspectives on Science and Prac-*

tice, 3(2), 154–8. See also Fiori, M., Antonietti, J.P., Mikolajczak, M., Luminet, O., Hansenne, M. and Rossier, J. (2014), "What Is the Ability Emotional Intelligence Test (MSCEIT) Good For? An Evaluation Using Item Response Theory," *PLOS One*, 9(6), e98827.

41 See Waterhouse, L. (2006), "Multiple Intelligences, the Mozart Effect, and Emotional Intelligence: A Critical Review," *Educational Psychologist*, 41(4), 207–25. And Waterhouse, L. (2006), "Inadequate Evidence for Multiple Intelligences, Mozart Effect, and Emotional Intelligence Theories," *Educational Psychologist*, 41(4), 247–55.

42 In the following paper, Robert Sternberg contrasts his theories to multiple intelligences and EQ: Sternberg, R.J. (1999), "Successful Intelligence: Finding a Balance," *Trends in Cognitive Sciences*, 3(11), 436–42.

43 Hagbloom, S.J., et al. (2002), "The 100 Most Eminent Psychologists of the 20th Century," *Review of General Psychology*, 6(2), 139–52.

44 Sternberg describes this journey in more detail in the following web post: http://www.cdl.org/articles/the-teachers-we-never-forget/.

45 For a more in-depth discussion, see Sternberg, R.J., and Preiss, D.D. (eds) (2010), *Innovations in Educational Psychology: Perspectives on Learning, Teaching, and Human Development*. New York: Springer, pp. 406–7.

46 Sternberg, "Successful Intelligence."

47 See, for instance, Hedlund, J., Wilt, J.M., Nebel, K.L., Ashford, S.J., and Sternberg, R.J. (2006), "Assessing Practical Intelligence in Business School Admissions: A Supplement to the Graduate Management Admissions Test," *Learning and Individual Differences*, 16(2), 101–27.

48 See the following PBS interview with Sternberg for a more in-depth discussion of these ideas: https://www.pbs.org/wgbh/pages/frontline/shows/sats/interviews/sternberg.html.

49 For summaries of these results, see Sternberg, R.J., Castejón, J.L., Prieto, M.D., Hautamäki, J., and Grigorenko, E.L. (2001), "Confirmatory Factor Analysis of the Sternberg Triarchic Abilities Test in Three International Samples: An Empirical Test of the Triarchic Theory of Intelligence," *European Journal of Psychological Assessment*, 17(1), 1–16. Sternberg, R.J. (2015), "Successful Intelligence: A Model for Testing Intelligence beyond IQ Tests," *European Journal of Education and Psychology*, 8(2), 76–84. Sternberg, R.J. (2008), "Increasing Academic Excellence and Enhancing Diversity Are Compatible Goals," *Educational Policy*, 22(4), 487–514. Sternberg, R.J., Grigorenko, E.L., and Zhang, L.F. (2008), "Styles of Learning and Thinking Matter in Instruction and Assessment," *Perspectives on Psychological Science*, 3(6), 486–506.

50 Sternberg, R.J. (2000), *Practical Intelligence in Everyday Life*, Cambridge: Cambridge University Press, pp. 144–200. See also Wagner, R.K., and Sternberg, R.J. (1985), "Practical Intelligence in Real-World Pursuits: The Role of Tacit Knowledge," *Journal of Personality and Social Psychology*, 49(2), 436–58. See also Cianciolo, A.T., et al. (2006), "Tacit Knowledge, Practical Intelligence and Expertise," in Ericsson, K.A.

(ed.), *Cambridge Handbook of Expertise and Expert Performance*, Cambridge: Cambridge University Press. For an independent discussion of Sternberg's studies, see Perkins, D. (1995), *Outsmarting IQ: The Emerging Science of Learnable Intelligence*, New York: Free Press, pp. 83–4. And Nisbett, R.E., Aronson, J., Blair, C., Dickens, W., Flynn, J., Halpern, D.F., and Turkheimer, E. (2012), "Intelligence: New Findings and Theoretical Developments," *American Psychologist*, 67(2), 130. And Mackintosh, N.J. (2011), *IQ and Human Intelligence*, Oxford: Oxford University Press, pp. 222–43.

51 In 1996, the APA's comprehensive report on "Intelligence: Known and Unknowns" concluded that "although this work is not without its critics, the results to this point tend to support the distinction between analytic and practical intelligence." Neisser et al., "Intelligence."

52 See, for instance: Imai, L., and Gelfand, M.J. (2010), "The Culturally Intelligent Negotiator: The Impact of Cultural Intelligence (CQ) on Negotiation Sequences and Outcomes," *Organizational Behavior and Human Decision Processes*, 112(2), 83–98. Alon, I., and Higgins, J.M. (2005), "Global Leadership Success through Emotional and Cultural Intelligences," *Business Horizons*, 48(6), 501–12. Rockstuhl, T., Seiler, S., Ang, S., Van Dyne, L., and Annen, H. (2011), "Beyond General Intelligence (IQ) and Emotional Intelligence (EQ): The Role of Cultural Intelligence (CQ) on Cross-Border Leadership Effectiveness in a Globalized World," *Journal of Social Issues*, 67(4), 825–40.

53 Marks, R. (2007), "Lewis M. Terman: Individual Differences and the Construction of Social Reality," *Educational Theory*, 24(4), 336–55.

54 Terman, L.M. (1916), *The Measurement of Intelligence: An Explanation of and a Complete Guide for the Use of the Stanford Revision and Extension of the Binet-Simon Intelligence Scale*. Boston, MA: Houghton Mifflin.

55 Lippmann, W. (25 October 1922), "The Mental Age of Americans," *New Republic*, p. 213.

56 Terman, L.M. (27 December 1922), "The Great Conspiracy or the Impulse Imperious of Intelligence Testers, Psychoanalyzed and Exposed by Mr. Lippmann," *New Republic*, p. 116.

57 Shurkin, *Terman's Kids*, p. 190.

58 Minton, H.L. (1988), *Lewis M. Terman: Pioneer in Psychological Testing*, New York, New York University Press.

Chapter 2

1 These passages draw on the following material: Ernst, B.M.L., and Carrington, H. (1933), *Houdini and Conan Doyle: The Story of a Strange Friendship*, London: Hutchinson. Conan Doyle, A.C. (1930), *The Edge of the Unknown*, London: John Murray. Kalush, W., and Sloman, L. (2006), *The Secret Life of Houdini: The Making of America's First Superhero*, New York: Atria. Sandford, C. (2011), *Houdini and Conan Doyle*, London: Duckworth Overlook. Gardner, L. (10 August 2015), "Harry Houdini and Arthur Conan Doyle: A Friendship Split by Spiritualism," *Guardian*, https://www.theguardian.com/stage/2015/aug/10/houdini-and-conan-Conan Doyle-impossible-edinburgh-festival.

2 Wilk, T. (2 May 2012), "Houdini, Sir Doyle Do AC," *Atlantic City Weekly*, http://www.atlanticcityweekly.com/news_and_views/houdini-sir-doyle-do-ac/article_a16ab3ba-95b9-50e1-a2e0-eca01dd8eaae.html.

3 In *Of Miracles*, the eighteenth-century philosopher David Hume put it like this: "No testimony is sufficient to establish a miracle, unless the testimony be of such a kind, that its falsehood would be more miraculous, than the fact, which it endeavours to establish." In other words, an extraordinary claim requires extraordinary evidence that discounts any physical explanations.

4 Fox newsreel of an interview with Sir Arthur Conan Doyle (1927). Available at *Public Domain Review*, https://publicdomainreview.org/collections/sir-arthur-conan-doyle-interview-1927/.

5 Eby, M. (21 March 2012), "Hocus Pocus," *Paris Review* blog, https://www.theparisreview.org/blog/2012/03/21/hocus-pocus/.

6 Tversky, A., and Kahneman, D. (1974), "Judgment under Uncertainty: Heuristics and Biases," *Science*, 185, 1124–31.

7 For an accessible description of this argument, see Stanovich, K.E. (2009), "Rational and Irrational Thought: The Thinking That IQ Tests Miss," *Scientific American Mind*, 20(6), 34–9.

8 There is good evidence, for instance, that children naturally reject information if it contradicts "common sense" theories of the world, and they need to learn the scientific method from the people they trust. So a child growing up in an environment that rejects science will naturally adopt those views, regardless of their intelligence. Bloom, P., and Weisberg, D.S. (2007), "Childhood Origins of Adult Resistance to Science," *Science*, 316(5827), 996–7.

9 "Knowledge projection from an island of false beliefs might explain the phenomenon of otherwise intelligent people who get caught in a domain-specific web of falsity that, because of projection tendencies, they cannot escape. Such individuals often use their considerable computational power to rationalize their beliefs and to ward off the arguments of skeptics." Stanovich, K.E., West, R.F., and Toplak, M.E. (2016), *The Rationality Quotient: Toward a Test of Rational Thinking*, Cambridge, MA: MIT Press. Kindle Edition (Kindle Locations 3636–9).

10 Stanovich, K. (1993), "Dysrationalia: A New Specific Learning Difficulty," *Journal of Learning Difficulties*, 26(8), 501–15.

11 For a helpful explanation of the principles, see Swinscow, T.D.V. (1997), *Statisics at Square One*, ninth edition. Available online at http://www.bmj.com/about-bmj/resources-readers/publications/statistics-square-one/11-correlation-and-regression.

12 Stanovich, West, and Toplak, *The Rationality Quotient* (Kindle Locations 2757, 2838). Some early studies had suggested the correlations to be even weaker. See Stanovich, K.E., and West, R.F. (2008), "On the Relative Independence of Thinking Biases and Cognitive Ability," *Journal of Personality and Social Psychology*, 94(4), 672–95.

13 Stanovich and West, "On the Relative Independence of Thinking Biases and Cognitive Ability."

14 Xue, G., He, Q., Lei, X., Chen, C., Liu, Y., Chen, C., et al. (2012), "The Gambler's Fallacy Is Associated with Weak Affective Decision Making but Strong Cognitive Ability," *PLOS One*, 7(10): e47019, https://doi.org/10.1371/journal.pone.0047019.

15 Schwitzgebel, Eric, and Fiery Cushman (2015), "Philosophers" Biased Judgments Persist Despite Training, Expertise and Reflection," *Cognition*, 141, 127–37.

16 West, R.F., Meserve, R.J., and Stanovich, K.E. (2012), "Cognitive Sophistication Does Not Attenuate the Bias Blind Spot," *Journal of Personality and Social Psychology*, 103(3), 506–19.

17 Stanovich, West, and Toplak, *The Rationality Quotient.*

18 Stanovich, K.E., and West, R.F. (2014), "What Intelligence Tests Miss," *Psychologist*, 27, 80–3, https://thepsychologist.bps.org.uk/volume-27/edition-2/what-intelligence-tests-miss.

19 Stanovich, West, and Toplak, *The Rationality Quotient* (Kindle Location 2344).

20 Bruine de Bruin, W., Parker, A.M., and Fischhoff, B. (2007), "Individual Differences in Adult Decision Making Competence," *Journal of Personality and Social Psychology*, 92(5), 938–56.

21 Kanazawa, S., and Hellberg, J.E.E.U. (2010), "Intelligence and Substance Use," *Review of General Psychology*, 14(4), 382–96.

22 Zagorsky, J. (2007), "Do You Have to Be Smart to Be Rich? The Impact of IQ on Wealth, Income and Financial Distress," *Intelligence*, 35, 489–501.

23 Swann, M. (8 March 2013). The professor, the bikini model, and the suitcase full of trouble. *New York Times*. http://www.nytimes.com/2013/03/10/magazine/the-professor-the-bikini-model-and-the-suitcase-full-of-trouble.html.

24 Rice, T.W. (2003), "Believe It or Not: Religious and Other Paranormal Beliefs in the United States," *Journal for the Scientific Study of Religion*, 42(1), 95–106.

25 Bouvet, R., and Bonnefon, J.F. (2015), "Non-reflective Thinkers Are Predisposed to Attribute Supernatural Causation to Uncanny Experiences," *Personality and Social Psychology Bulletin*, 41(7), 955–61.

26 Cooper, J. (1990), *The Case of the Cottingley Fairies*, London: Robert Hale.

27 Conan Doyle, A.C. (1922), *The Coming of the Fairies*, London: Hodder & Stoughton.

28 Cooper, J. (1982), "Cottingley: At Last the Truth," *The Unexplained*, 117, 2338–40.

29 Miller, R. (2008), *The Adventures of Arthur Conan Doyle*, London: Harvill Secker, p. 403.

30 Hyman, R. (2002), in *Why Smart People Can Be So Stupid*, ed. Sternberg, R., New Haven: Yale University Press, pp. 18–19.

31 Perkins, D.N., Farady, M., and Bushey, B. (1991), "Everyday Reasoning and the Roots of Intelligence," in Perkins, D., Voss, James F., and Segal, Judith W. (eds), *Informal Reasoning and Education*, Hillsdale, NJ: Erlbaum, pp. 83–105.

32 For a fuller discussion of this study, see Perkins, D.N. (1995), *Outsmarting IQ*, New York: Free Press, pp. 131–5.

33 Perkins, D.N., and Tishman, S. (2001), "Dispositional Aspects of Intelligence," in

Collis, J.M. and Messick, S. (eds), *Intelligence and Personality: Bridging the Gap in Theory and Measurement*, Hillsdale, NJ: Erlbaum, pp. 233–57.

34 Kahan, D.M., Peters, E., Dawson, E.C., and Slovic, P. (2017), "Motivated Numeracy and Enlightened Self-Government," *Behavioural Public Policy*, 1, 54–86.

35 For a fuller discussion of the ways that increased knowledge can backfire, see Flynn, D.J., Nyhan, B., and Reifler, J. (2017), "The Nature and Origins of Misperceptions: Understanding False and Unsupported Beliefs about Politics," *Advances in Political Psychology*, 38(S1), 127–50. And also Taber, C.S. and Lodge, M. (2006), "Motivated Skepticism in the Evaluation of Political Beliefs," *American Journal of Political Science*, 50, 755–69.

36 Kahan, D.M., et al. (2012), "The Polarizing Impact of Science Literacy and Numeracy on Perceived Climate Change Risks," *Nature Climate Change*, 2(10), 732–5. Kahan, D.M., Wittlin, M., Peters, E., Slovic, P., Ouellette, L.L., Braman, D., and Mandel, G.N. (2011), "The Tragedy of the Risk-Perception Commons: Culture Conflict, Rationality Conflict, and Climate Change," https://papers.ssrn.com/sol3/papers.cfm?abstract_id=1871503. Bolsen, T., Druckman, J.N., and Cook, F.L. (2015), "Citizens', Scientists', and Policy Advisors' Beliefs about Global Warming," *Annals of the American Academy of Political and Social Science*, 658(1), 271–95.

37 Hamilton, L.C., Hartter, J., and Saito, K. (2015), "Trust in Scientists on Climate Change and Vaccines," *SAGE Open*, 5(3), doi: https://doi.org/10.1177/2158244015602752.

38 Kahan, D.M., Landrum, A., Carpenter, K., Helft, L., and Hall Jamieson, K. (2017), "Science Curiosity and Political Information Processing," *Political Psychology*, 38(S1), 179–99.

39 Drummond, C., and Fischhoff, B. (2017), "Individuals with greater science literacy and education have more polarized beliefs on controversial science topics," *Proceedings of the National Academy of Sciences*, 114(36), 9587–92.

40 Nyhan, B., Reifler, J., and Ubel, P.A. (2013), "The Hazards of Correcting Myths about Health Care Reform," *Medical Care*, 51(2), 127–32. For a discussion of the misperceptions around ObamaCare, see *Politifact*'s Lie of the Year: "Death Panels," *Politifact*, 18 December 2009, http://www.politifact.com/truth-o-meter/article/2009/dec/18/politifact-lie-year-death-panels/.

41 Koehler, J.J. (1993), "The Influence of Prior Beliefs on Scientific Judgments of Evidence Quality," *Organizational Behavior and Human Decision Processes*, 56(1), 28–55. See also Dan Kahan's discussion of the paper, in light of recent research on motivated reasoning: Kahan, D.M. (2016), "The Politically Motivated Reasoning Paradigm, Part 2: What Politically Motivated Reasoning Is and How to Measure It," in *Emerging Trends in the Social and Behavioral Sciences*, doi: 10.1002/9781118900772.

42 Including, apparently, $25,000 from his US book tour of 1922. Ernst, B.M.L., and Carrington, H. (1971), *Houdini and Conan Doyle: The Story of a Strange Friendship*, New York: Benjamin Blom, p. 147.

43 In this recording from the British Library, Conan Doyle explains the many benefits he has drawn from his belief in spiritualism: http://britishlibrary.typepad.co.uk/files/listen-to-sir-arthur-conan-doyle-on-spiritualism.mp3.

44 Fox newsreel of an interview with Sir Arthur Conan Doyle (1927). Available at *Public Domain Review*, https://publicdomainreview.org/collections/sir-arthur-conan-doyle-interview-1927/.

45 As his biographer, Russell Miller, describes: "Once Conan Doyle made up his mind, he was unstoppable, impervious to argument, blind to contradictory evidence, untroubled by self-doubt." Miller, *The Adventures of Arthur Conan Doyle*, chapter 20.

46 Bechtel, S., and Stains, L.R. (2017), *Through a Glass Darkly: Sir Arthur Conan Doyle and the Quest to Solve the Greatest Mystery of All*, New York: St Martin's Press, p. 147.

47 Panek, R. (2005), "The Year of Albert Einstein," *Smithsonian Magazine*, https://www.smithsonianmag.com/science-nature/the-year-of-albert-einstein-75841381/.

48 Further examples can be found in the following interview with the physicist John Moffat, including the fact that Einstein denied strong evidence for the existence of black holes: Folger, T. (September 2004), "Einstein's Grand Quest for a Unified Theory," *Discover*, http://discovermagazine.com/2004/sep/einsteins-grand-quest. See also Mackie, G. (2015), "Einstein's Folly: How the Search for a Unified Theory Stumped Him until His Dying Day," *The Conversation*, http://theconversation.com/einsteins-folly-how-the-search-for-a-unified-theory-stumped-him-to-his-dying-day-49646.

49 Isaacson, W. (2007), *Einstein: His Life and Universe*, London: Simon & Schuster, pp. 341–7.

50 Schweber, S.S. (2008), *Einstein and Oppenheimer: The Meaning of Genius*, Cambridge, MA: Harvard University Press, p. 282. See also Oppenheimer, R. (17 March 1966), "On Albert Einstein," *New York Review of Books*, http://www.nybooks.com/articles/1966/03/17/on-albert-einstein/.

51 Hook, S. (1987), *Out of Step: An Unquiet Life in the 20th Century*. London: Harper & Row. See also Riniolo, T., and Nisbet, L. (2007), "The Myth of Consistent Skepticism: The Cautionary Case of Albert Einstein," *Skeptical Inquirer*, 31(1), http://www.csicop.org/si/show/myth_of_consistent_skepticism_the_cautionary_case_of_albert_einstein.

52 Eysenck, H. (1957), *Sense and Nonsense in Psychology*, Harmondsworth, UK: Penguin, p. 108.

53 These are exceptional cases. But the issue of bias in the day-to-day workings of science has come to increasing prominence in recent years, with concerns that many scientists may engage in the wishful thinking that had plagued Conan Doyle. In the 1990s and early 2000s, the psychologist Kevin Dunbar spent years studying the thinking of scientists at eight different laboratories—attending their weekly meetings and discussing their latest findings. He found that myside thinking was rife, with many scientists unconsciously distorting the interpretation of

their experimental results to fit their current hypothesis, or deliberately searching for new and more convoluted reasons to make their hypotheses fit the data. Medical researchers appear particularly likely to cling to newsworthy results, while overlooking serious methodological flaws. See, for instance, Dunbar, K. (2000), "How Scientists Think in the Real World," *Journal of Applied Developmental Psychology*, 21(1), 49–58. Wilson, T.D., DePaulo, B.M., Mook, D.G., and Klaaren, K.J. (1993), "Scientists' Evaluations of Research: The Biasing Effects of the Importance of the Topic," *Psychological Science*, 4(5), 322–5.

54 Offit, P. (2013), "The Vitamin Myth: Why We Think We Need Supplements," *The Atlantic*, 19 July 2013, https://www.theatlantic.com/health/archive/2013/07/the-vitamin-myth-why-we-think-we-need-supplements/277947/.

55 Enserink, M. (2010), "French Nobelist Escapes 'Intellectual Terror' To Pursue Radical Ideas In China," *Science*, 330(6012), 1732. For a further discussion of these controversies, see: Butler, D. (2012). Nobel fight over African HIV center. *Nature*, 486(7403), 301-2. https://www.nature.com/news/nobel-fight-over-african-hiv-centre-1.10847

56 King, G. (2011), "Edison vs. Westinghouse: A Shocking Rivalry," *Smithsonian Magazine*, http://www.smithsonianmag.com/history/edison-vs-westinghouse-a-shocking-rivalry-102146036/.

57 Essig, M. (2003), *Edison and the Electric Chair*, Stroud, Gloucestershire: Sutton.

58 Essig, *Edison and the Electric Chair*, p. 289.

59 Essig, *Edison and the Electric Chair*, p. 289.

60 Isaacson, W. (2011), *Steve Jobs*, London: Little, Brown, pp. 422–55. Swaine, J. (21 October 2011), "Steve Jobs 'Regretted Trying to Beat Cancer with Alternative Medicine for So Long'," *Daily Telegraph*, http://www.telegraph.co.uk/technology/apple/8841347/Steve-Jobs-regretted-trying-to-beat-cancer-with-alternative-medicine-for-so-long.html.

61 Shultz, S., Nelson, E. and Dunbar, R.I.M. (2012), "Hominin Cognitive Evolution: Identifying Patterns and Processes in the Fossil and Archaeological Record," *Philosophical Transactions of the Royal Society B: Biological Sciences*, 367(1599), 2130–40.

62 Mercier, H. (2016), "The Argumentative Theory: Predictions and Empirical Evidence," *Trends in Cognitive Sciences*, 20(9), 689–700.

Chapter 3

1 The details of Brandon Mayfield's experiences have been taken from my own interviews, as well as his press interviews, including a video interview with Open Democracy (30 November 2006) at https://www.democracynow.org/2006/11/30/exclusive_falsely_jailed_attorney_brandon_mayfield. I am also indebted to Mayfield, S. and Mayfield, B. (2015), *Improbable Cause: The War on Terror's Assault on the Bill of Rights*, Salem, NH: Divertir. I have cross-checked many of the details with the Office of the Inspector General's report on the FBI's handling of Mayfield's case.

2 Jennifer Mnookin of UCLA, in "Fingerprints on Trial," BBC World Service, 29 March 2011, http://www.bbc.co.uk/programmes/poofvhl3.

3 Office of the Inspector General (2006), "A Review of the FBI's Handling of the Brandon Mayfield Case," p. 80, https://oig.justice.gov/special/s0601/final.pdf.

4 Office of the Inspector General, "A Review of the FBI's Handling of the Brandon Mayfield Case," p. 80.

5 Kassin, S.M., Dror, I.E., and Kukucka, J. (2013), "The Forensic Confirmation Bias: Problems, Perspectives, and Proposed Solutions," *Journal of Applied Research in Memory and Cognition*, 2(1), 42–52.

6 Fisher, R. (2011), "Erudition Be Damned, Ignorance Really Is Bliss," *New Scientist*, 211(2823), 39–41.

7 Kruger, J., and Dunning, D. (1999), "Unskilled and Unaware of It: How Difficulties in Recognizing One's Own Incompetence Lead to Inflated Self-assessments," *Journal of Personality and Social Psychology*, 77(6), 1121–34.

8 Dunning, D. (2011), "The Dunning–Kruger Effect: On Being Ignorant of One's Own Ignorance," in *Advances in Experimental Social Psychology*, Vol. 44, Cambridge, MA: Academic Press, pp. 247–96.

9 Chiu, M.M., and Klassen, R.M. (2010), "Relations of Mathematics Self-Concept and Its Calibration with Mathematics Achievement: Cultural Differences among Fifteen-Year-Olds in 34 Countries," *Learning and Instruction*, 20(1), 2–17.

10 See, for example, "Why Losers Have Delusions of Grandeur," *New York Post*, 23 May 2010, https://nypost.com/2010/05/23/why-losers-have-delusions-of-grandeur/. Lee, C. (2016), "Revisiting Why Incompetents Think They Are Awesome," *Ars Technica*, 4 November 2016, https://arstechnica.com/science/2016/11/revisiting-why-incompetents-think-theyre-awesome/. Flam, F. (2017), "Trump's 'Dangerous Disability'? The Dunning–Kruger Effect," Bloomberg, 12 May 2017, https://www.bloomberg.com/view/articles/2017-05-12/trump-s-dangerous-disability-it-s-the-dunning-kruger-effect.

11 Fisher, M., and Keil, F.C. (2016), "The Curse of Expertise: When More Knowledge Leads to Miscalibrated Explanatory Insight," *Cognitive Science*, 40(5), 1251–69.

12 Son, L.K., and Kornell, N. (2010), "The Virtues of Ignorance," *Behavioural Processes*, 83(2), 207–12.

13 Fisher and Keil, "The Curse of Expertise."

14 Ottati, V., Price, E., Wilson, C., and Sumaktoyo, N. (2015), "When Self-Perceptions of Expertise Increase Closed-Minded Cognition: The Earned Dogmatism Effect," *Journal of Experimental Social Psychology*, 61, 131–8.

15 Quoted in Hammond, A.L. (1984). *A Passion to Know: Twenty Profiles in Science*, New York: Scribner, p. 5. This viewpoint is also discussed, in depth, in Roberts, R.C., Wood, W.J. (2007), *Intellectual Virtues: An Essay in Regulative Epistemology*, Oxford: Oxford University Press, p. 253.

16 Much of this information on de Groot's life comes from an obituary in the *Observer* of the American Psychological Society, published online on 1 November 2006: http://www. psychologicalscience.org/observer/in-memoriam-adriaan-dingeman-de-groot-1914-2006#.WUpLDIrTUdV.

17 Mellenbergh, G.J., and Hofstee, W.K.B. (2006), "Commemoration Adriaan Dingeman de Groot," in: Royal Netherlands Academy of Sciences (ed.), *Life and Memorials*, Amsterdam: Royal Netherlands Academy of Sciences, pp. 27–30.

18 Busato, V. (2006), "In Memoriam: Adriaan de Groot (1914–2006)," *Netherlands Journal of Psychology*, 62, 2–4.

19 de Groot, A. (ed.) (2008), *Thought and Choice in Chess*, Amsterdam: Amsterdam University Press, p. 288. See also William Chase and Herbert Simon's classic follow-up experiment that provides further evidence for the role of chunking in expert performance: Chase, W. G., and Simon, H. A. (1973). "Perception in Chess." Cognitive Psychology, 4(1), 55-81.

20 Hodges, N.J., Starkes, J.L., and MacMahon, C. (2006), "Expert Performance in Sport: A Cognitive Perspective," in Ericsson, K.A., Charness, N., Feltovich, P.J., et al. (eds), *Cambridge Handbook of Expertise and Expert Performance*, Cambridge: Cambridge University Press.

21 Dobbs, D. (2006), "How to Be a Genius," *New Scientist*, 191(2569), 40–3.

22 Kalakoski, V. and Saariluoma, P. (2001), "Taxi Drivers' Exceptional Memory of Street Names," *Memory and Cognition*, 29(4), 634–8.

23 Nee, C., and Ward, T. (2015), "Review of Expertise and Its General Implications for Correctional Psychology and Criminology," *Aggression and Violent Behavior*, 20, 1–9.

24 Nee, C., and Meenaghan, A. (2006), "Expert Decision Making in Burglars," *British Journal of Criminology*, 46(5), 935–49.

25 For a comprehensive review of the evidence, see Dane, E. (2010), "Reconsidering the Trade-Off between Expertise and Flexibility: A Cognitive Entrenchment Perspective," *Academy of Management Review*, 35(4), 579–603.

26 Woollett, K., and Maguire, E.A. (2010), "The Effect of Navigational Expertise on Wayfinding in New Environments," *Journal of Environmental Psychology*, 30(4), 565–73.

27 Harley, E.M., Pope, W.B., Villablanca, J.P., Mumford, J., Suh, R., Mazziotta, J.C., Enzmann, D., and Engel, S.A. (2009), "Engagement of Fusiform Cortex and Disengagement of Lateral Occipital Cortex in the Acquisition of Radiological Expertise," *Cerebral Cortex*, 19(11), 2746–54.

28 See, for instance: Corbin, J.C., Reyna, V.F., Weldon, R.B., and Brainerd, C.J. (2015), "How Reasoning, Judgment, and Decision Making Are Colored by Gist-Based Intuition: A Fuzzy-Trace Theory Approach," *Journal of Applied Research in Memory and Cognition*, 4(4), 344–55. The following chapter also gives a more complete description of many of the findings in the preceding paragraphs: Dror, I.E. (2011), "The Paradox of Human Expertise: Why Experts Get It Wrong," in *The Paradoxical Brain*, ed. Narinder Kapur, Cambridge: Cambridge University Press.

29 Northcraft, G.B., and Neale, M.A. (1987), "Experts, Amateurs, and Real Estate: An Anchoring-and-Adjustment Perspective on Property Pricing Decisions," *Organizational Behavior and Human Decision Processes*, 39(1), 84–97.

30 Busey, T.A., and Parada, F.J. (2010), "The Nature of Expertise in Fingerprint Examiners," *Psychonomic Bulletin and Review*, 17(2), 155–60.

31 Busey, T.A., and Vanderkolk, J.R. (2005), "Behavioral and Electrophysiological Evidence for Configural Processing in Fingerprint Experts," *Vision Research*, 45(4), 431–48.

32 Dror, I.E. and Charlton, D. (2006), "Why Experts Make Errors," *Journal of Forensic Identification*, 56(4), 600–16.

33 Dror, I.E., Péron, A.E., Hind, S.-L., and Charlton, D. (2005), "When Emotions Get the Better of Us: The Effect of Contextual Top-Down Processing on Matching Fingerprints," *Applied Cognitive Psychology*, 19(6), 799–809.

34 Office of the Inspector General, "A Review of the FBI's Handling of the Brandon Mayfield Case," p. 192.

35 Office of the Inspector General, "A Review of the FBI's Handling of the Brandon Mayfield Case," p. 164.

36 Dror, I.E., Morgan, R., Rando, C., and Nakhaeizadeh, S. (2017), "The Bias Snowball and the Bias Cascade Effects: Two Distinct Biases That May Impact Forensic Decision Making," *Journal of Forensic Science*, 62(3), 832–3.

37 Office of the Inspector General, "A Review of the FBI's Handling of the Brandon Mayfield Case," p. 179.

38 Kershaw, S. (5 June 2004), "Spain at Odds on Mistaken Terror Arrest," *New York Times*, http://www.nytimes.com/learning/students/pop/articles/05LAWY.html.

39 Office of the Inspector General, "A Review of the FBI's Handling of the Brandon Mayfield Case," p. 52.

40 Dismukes, K., Berman, B.A., and Loukopoulos, L.D. (2007), *The Limits of Expertise: Rethinking Pilot Error and the Causes of Airline Accidents*, Aldershot: Ashgate, pp. 76–81.

41 National Transport Safety Board [NTSB] (2008), "Attempted Takeoff from Wrong Runway Comair Flight 5191, 27 August 2006," Accident Report NTSB/AAR-07/05. This report specifically cites confirmation bias—the kind explored by Stephen Walmsley and Andrew Gilbey as one of the primary sources of the error—and Walmsley and Gilbey cite it as an inspiration for their paper.

42 Walmsley, S., and Gilbey, A. (2016), "Cognitive Biases in Visual Pilots' Weather-Related Decision Making," *Applied Cognitive Psychology*, 30(4), 532–43.

43 Levinthal, D., and Rerup, C. (2006), "Crossing an Apparent Chasm: Bridging Mindful and Less-Mindful Perspectives on Organizational Learning," *Organization Science*, 17(4), 502–13.

44 Kirkpatrick, G. (2009), "The Corporate Governance Lessons from the Financial Crisis," *OECD Journal: Financial Market Trends*, 2009(1), 61–87.

45 Minton, B. A., Taillard, J. P., and Williamson, R. (2014). "Financial Expertise of the Board, Risk Taking, and Performance: Evidence from Bank Holding Companies." *Journal of Financial and Quantitative Analysis*, 49(2), 351-380. Juan Almandoz at the IESE Business School in Barcelona and András Tilcsik at the University of Toronto have found the same patterns in the board members and CEOs of local banks in the United States. Like Williamson, they found that the more experts the banks had on their board, the more likely they were to fail during times of uncertainty, due to entrenchment, overconfidence, and the suppression of alternative ideas. Almandoz, J., and Tilcsik, A. (2016), "When Experts Become Liabilities: Domain Experts on Boards and Organizational Failure," *Academy of Management Journal*, 59(4), 1124–49. Monika Czerwonka at the Warsaw School of Economics, meanwhile, has found that expert stock market investors are more susceptible to the sunk cost bias—the tendency to pour more money into a failing investment, even if we know it is making a loss. Again, the greater their expertise, the greater their vulnerability. Rzeszutek, M., Szyszka, A., and Czerwonka, M. (2015), "Investors' Expertise, Personality Traits and Susceptibility to Behavioral Biases in the Decision Making Process," *Contemporary Economics*, 9, 337–52.

46 Jennifer Mnookin of UCLA, in *Fingerprints on Trial*, BBC World Service, 29 March 2011, http://www.bbc.co.uk/programmes/poofvhl3.

47 Dror, I.E., Thompson, W.C., Meissner, C.A., Kornfield, I., Krane, D., Saks, M., and Risinger, M. (2015), "Letter to the Editor—Context Management Toolbox: A Linear Sequential Unmasking (LSU) Approach for Minimizing Cognitive Bias in Forensic Decision Making," *Journal of Forensic Sciences*, 60(4), 1111–12.

Chapter 4

1 Brown, B. (2012), "Hot, Hot, Hot: The Summer of 1787," National Constitution Center blog, https://constitutioncenter.org/blog/hot-hot-hot-the-summer-of-1787.

2 For much of the background detail on the US Constitution I am indebted to Isaacson, W. (2003), *Benjamin Franklin: An American Life*, New York: Simon & Schuster.

3 Franklin, B. (19 April 1787), Letter to Thomas Jefferson, Philadelphia. Retrieved from Franklin's online archives, courtesy of the American Philosophical Society and Yale University, http://franklinpapers.org/franklin/framedVolumes.jsp?vol=44&page=613.

4 Madison Debates (30 June 1787). Retrieved from the Avalon project at Yale Law School, http://avalon.law.yale.edu/18th_century/debates_630.asp.

5 Madison Debates (17 September 1787). Retrieved from the Avalon project at Yale Law School, http://avalon.law.yale.edu/18th_century/debates_917.asp.

6 Isaacson, W. (2003), *Benjamin Franklin: An American Life*, New York: Simon & Schuster, p. 149.

7 Lynch, T.J., Boyer, P.S., Nichols, C., and Milne, D. (2013), *The Oxford Encyclopedia of American Military and Diplomatic History*, New York: Oxford University Press, p. 398.

8 Proceedings from Benjamin Franklin's debating club define wisdom as "The Knowledge of what will be best for us on all Occasions and of the best Ways of attaining it." They also declared that no man is "wise at all Times or in all Things" though "some are much more frequently wise than others" (*Proposals and Queries to be Asked the Junto*, 1732).

9 "Conversations on Wisdom: UnCut Interview with Valerie Tiberius," from the Chicago Center for Practical Wisdom, https://www.youtube.com/watch?v=oFu

ToyY2otw. See also Tiberius, V. (2016), "Wisdom and Humility," *Annals of the New York Academy of Sciences*, 1384, 113–16.

10 Birren, J.E., and Svensson, C.M. (2005), in Sternberg, R. and Jordan, J. (eds), *A Handbook of Wisdom: Psychological Perspectives*, Cambridge: Cambridge University Press, pp. 12–13. As Birren and Svensson point out, early psychologists preferred to look at "psychophysics"—exploring, for instance, the basic elements of perception—and they would have considered wisdom too complex to pin down in a laboratory. And the topic tended to be shunned until well into the twentieth century, with a notable absence in many of the major textbooks, including *An Intellectual History of Psychology* (Daniel Robinson, 1976) and the *Handbook of General Psychology* (Benjamin Wolman, 1973).

11 Sternberg, R.J., Bonney, C.R., Gabora, L., and Merrifield, M. (2012), "WICS: A Model for College and University Admissions," *Educational Psychologist*, 47(1), 30–41.

12 Grossmann, I. (2017), "Wisdom in Context," *Perspectives on Psychological Science*, 12(2), 233–57.

13 Grossmann, I., Na, J., Varnum, M.E.W., Kitayama, S., and Nisbett, R.E. (2013), "A Route to Well-Being: Intelligence vs. Wise Reasoning," *Journal of Experimental Psychology. General*, 142(3), 944–53.

14 Bruine de Bruin, W., Parker A.M., and Fischhoff B. (2007), "Individual Differences in Adult Decision Making Competence," *Journal of Personality and Social Psychology*, 92(5), 938–56.

15 Stanovich, K.E.E., West, R.F., and Toplak, M. (2016), *The Rationality Quotient*, Cambridge, MA: MIT Press. Along similar lines, various studies have shown that open-minded reasoning—an important component of Grossmann's definition of wisdom—leads to greater well-being and happiness. It also seems to make people more inquisitive about potential health risks: Lambie, J. (2014), *How to Be Critically Open-Minded: A Psychological and Historical Analysis*, Basingstoke: Palgrave Macmillan, pp. 89–90.

16 See http://wici.ca/new/2016/06/igor-grossmann/.

17 At the time of writing this was a preprint of the paper, awaiting peer review and publication. Santos, H.C., and Grossmann, I. (2018), "Relationship of Wisdom-Related Attitudes and Subjective Well-Being over Twenty Years: Application of the Train-Preregister-Test (TPT) Cross-Validation Approach to Longitudinal Data." Available at https://psyarxiv.com/f4thj/.

18 Grossmann, I., Gerlach, T.M., and Denissen, J.J.A. (2016), "Wise Reasoning in the Face of Everyday Life Challenges," *Social Psychological and Personality Science*, 7(7), 611–22.

19 Franklin, B. (1909), *The Autobiography of Benjamin Franklin*, p. 17. Public domain ebook of the 1909 Collier & Son edition.

20 Letter from Benjamin Franklin to John Lining (18 March 1775). Retrieved from the website of the US National Archives, https://founders.archives.gov/documents/Franklin/01-05-02-0149.

21 Lord, C.G., Ross, L., and Lepper, M.R. (1979), "Biased Assimilation and Attitude Polarization: The Effects of Prior Theories on Subsequently Considered Evidence," *Journal of Personality and Social Psychology*, 37(11), 2098–2109. Thanks to Tom Stafford for pointing me towards this paper and its interpretation, in his article for BBC Future: Stafford, T. (2017), "How to Get People to Overcome Their Bias," http://www.bbc.com/future/story/20170131-why-wont-some-people-listen-to-reason.

22 Isaacson, W. (2003), "*A Benjamin Franklin Reader*," p. 236. New York: Simon & Schuster.

23 Letter from Benjamin Franklin to Jonathan Williams, Jr. (8 April 1779). Retrieved from Franklin's online archives, courtesy of the American Philosophical Society and Yale University, http://franklinpapers.org/franklin//framedVolumes.jsp?vol=29&page=283a.

24 Jonas, E., Schulz-Hardt, S., Frey, D., and Thelen, N. (2001), "Confirmation Bias in Sequential Information Search after Preliminary Decisions: An Expansion of Dissonance Theoretical Research on Selective Exposure to Information," *Journal of Personality and Social Psychology*, 80(4), 557–71. You can also find a discussion of this paper, and its implications for decision making, in Church, I. (2016), *Intellectual Humility: An Introduction to the Philosophy and Science*, Bloomsbury. Kindle Edition (Kindle Locations 5817–20).

25 Baron, J., Gürçay, B., and Metz, S.E. (2016), "Reflection, Intuition, and Actively Open-Minded Thinking," in Weller, J. and Toplak, M.E. (eds), *Individual Differences in Judgment and Decision Making from a Developmental Context*, London: Psychology Press.

26 Adame, B.J. (2016), "Training in the Mitigation of Anchoring Bias: A Test of the Consider-the-Opposite Strategy," *Learning and Motivation*, 53, 36–48.

27 Hirt, E.R., Kardes, F.R., and Markman, K.D. (2004), "Activating a Mental Simulation Mind-Set Through Generation of Alternatives: Implications for Debiasing in Related and Unrelated Domains," *Journal of Experimental Social Psychology*, 40(3), 374–83.

28 Chandon, P., and Wansink, B. (2007), "The Biasing Health Halos of Fast-Food Restaurant Health Claims: Lower Calorie Estimates and Higher Side-Dish Consumption Intentions," *Journal of Consumer Research*, 34(3), 301–14.

29 Miller, A.K., Markman, K.D., Wagner, M.M., and Hunt, A.N. (2013), "Mental Simulation and Sexual Prejudice Reduction: The Debiasing Role of Counterfactual Thinking," *Journal of Applied Social Psychology*, 43(1), 190–4.

30 For a thorough examination of the "consider the opposite strategy" and its psychological benefits, see Lambie, J. (2014), *How to Be Critically Open-Minded: A Psychological and Historical Analysis*, Basingstoke: Palgrave Macmillan, pp. 82–6.

31 Herzog, S.M., and Hertwig, R. (2009), "The Wisdom of Many in One Mind: Improving Individual Judgments with Dialectical Bootstrapping," *Psychological Science*, 20(2), 231–7.

32 See the following paper for a recent examination of this technique: Fisher, M. and Keil, F.C. (2014), "The Illusion of Argument Justification," *Journal of Experimental Psychology: General*, 143(1), 425–33.

33 See the following paper for a review of this research: Samuelson, P.L., and Church, I.M. (2015), "When Cognition Turns Vicious: Heuristics and Biases in Light of Virtue Epistemology," *Philosophical Psychology*, 28(8), 1095–1113. The following paper provides an explanation of the reasons accountability may fail, if we do not feel comfortable enough to share the sources of our reasoning honestly: Mercier, H., Boudry, M., Paglieri, F., and Trouche, E. (2017), "Natural-born Arguers: Teaching How to Make the Best of Our Reasoning Abilities," *Educational Psychologist*, 52(1), 1–16.

34 Middlekauf, R. (1996), *Benjamin Franklin and His Enemies*, Berkeley: University of California Press, p. 57.

35 Suedfeld, P., Tetlock, P.E., and Ramirez, C. (1977), "War, Peace, and Integrative Complexity: UN Speeches on the Middle East Problem, 1947–1976," *Journal of Conflict Resolution*, 21(3), 427–42.

36 The psychologist John Lambie offers a more detailed analysis of these political and military studies in Lambie, *How to be Critically Open-Minded*, pp. 193–7.

37 See, for example, the following article by Patricia Hogwood, a reader in European Politics at the University of Westminster: Hogwood, P. (21 September 2017), "The Angela Merkel Model—or How to Succeed in German Politics," *The Conversation*, https://theconversation.com/the-angela-merkel-model-or-how-to-succeed-in-german-politics-84442

38 Packer, G. (1 December 2014), "The Quiet German: The Astonishing Rise of Angela Merkel, the Most Powerful Woman in the World," *New Yorker*: https://www.newyorker.com/magazine/2014/12/01/quiet-german.

39 Parker, K.I. (1992), "Solomon as Philosopher King? The Nexus of Law and Wisdom in 1 Kings 1–11," *Journal for the Study of the Old Testament*, 17(53), 75–91. Additional details drawn from Hirsch, E.G., et al. (1906), "Solomon," *Jewish Encyclopedia*. Retrieved online at: http://www.jewishencyclopedia.com/articles/13842-solomon.

40 Grossmann, I., and Kross, E. (2014), "Exploring Solomon's Paradox: Self-Distancing Eliminates the Self–Other Asymmetry in Wise Reasoning about Close Relationships in Younger and Older Adults," *Psychological Science*, 25(8), 1571–80.

41 Kross, E., Ayduk, O., and Mischel, W. (2005), "When Asking 'Why' Does Not Hurt: Distinguishing Rumination from Reflective Processing of Negative Emotions," *Psychological Science*, 16(9), 709–15.

42 For a wide-ranging review of this research, see Kross, E., and Ayduk, O. (2017), "Self-distancing," *Advances in Experimental Social Psychology*, 55, 81–136.

43 Streamer, L., Seery, M.D., Kondrak, C.L., Lamarche V.M., and Saltsman, T.L. (2017), "Not I, But She: The Beneficial Effects of Self-Distancing on Challenge/Threat Cardiovascular Responses," *Journal of Experimental Social Psychology*, 70, 235–41.

44 Grossmann and Kross, "Exploring Solomon's Paradox."

45 Finkel, E.J., Slotter, E.B., Luchies, L.B., Walton, G.M. and Gross, J.J. (2013), "A Brief Intervention to Promote Conflict Reappraisal Preserves Marital Quality Over Time," *Psychological Science*, 24(8), 1595–1601.

46 Kross, E., and Grossmann, I. (2012), "Boosting Wisdom: Distance from the Self Enhances Wise Reasoning, Attitudes, and Behavior," *Journal of Experimental Psychology: General*, 141(1), 43–8.

47 Grossmann, I. (2017), "Wisdom and How to Cultivate It: Review of Emerging Evidence for a Constructivist Model of Wise Thinking," *European Psychologist*, 22(4), 233–246.

48 Reyna, V.F., Chick, C.F., Corbin, J.C., and Hsia, A.N. (2013), "Developmental Reversals in Risky Decision Making: Intelligence Agents Show Larger Decision Biases than College Students," *Psychological Science*, 25(1), 76–84.

49 See, for example, Maddux, W.W., Bivolaru, E., Hafenbrack, A.C., Tadmor, C.T., and Galinsky, A.D. (2014), "Expanding Opportunities by Opening Your Mind: Multicultural Engagement Predicts Job Market Success through Longitudinal Increases in Integrative Complexity," *Social Psychological and Personality Science*, 5(5), 608–15.

50 Tetlock, P., and Gardner, D. (2015), *Superforecasting: The Art and Science of Prediction*, p. 126. London: Random House.

51 Grossmann, I., Karasawa, M., Izumi, S., Na, J., Varnum, M.E.W., Kitayama, S., and Nisbett, R. (2012), "Aging and Wisdom: Culture Matters," *Psychological Science*, 23(10), 1059–66.

52 Manuelo, E., Kusumi, T., Koyasu, M., Michita, Y., and Tanaka, Y. (2015), in Davies, M., and Barnett, R. (eds), *The Palgrave Handbook of Critical Thinking in Higher Education*, New York: Palgrave Macmillan, pp. 299–315.

53 For a review of this evidence, see Nisbett, R.E., Peng, K., Choi, I., and Norenzayan, A. (2001), "Culture and Systems of Thought: Holistic Versus Analytic Cognition," *Psychological Review*, 108(2), 291–310. Markus, H.R., and Kitayama, S. (1991), "Culture and the Self: Implications for Cognition, Emotion, and Motivation," *Psychological Review*, 98(2), 224–53. Henrich, J., Heine, S.J., and Norenzayan, A. (2010), "Beyond WEIRD: Towards a Broad-based Behavioral Science," *Behavioral and Brain Sciences*, 33(2–3), 111–35.

54 For more information on the sense of selfhood in Japan (and, in particular, the way it is encoded in language and education), see Cave, P. (2007), *Primary School in Japan: Self, Individuality, and Learning in Elementary Education*, Abingdon, England: Routledge, pp. 31–43, Smith, R. (1983), *Japanese Society: Tradition, Self, and the Social Order*, Cambridge: Cambridge University Press, pp. 68–105.

55 See, for example, Talhelm, T., Zhang, X., Oishi, S., Shimin, C., Duan, D., Lan, X., and Kitayama, S. (2014), "Large-scale Psychological Differences within China Explained by Rice versus Wheat Agriculture," *Science*, 344(6184), 603–8.

56 Henrich, Heine, and Norenzayan, "Beyond WEIRD."

57 See, for instance, Grossmann, I., and Kross, E. (2010), "The Impact of Culture on Adaptive versus Maladaptive Self-Reflection," *Psychological Science*, 21(8), 1150–7. Wu, S. and Keysar, B. (2007), "The Effect of Culture on Perspective Taking," *Psychological*

Science, 18(7), 600–6. Spencer-Rodgers, J., Williams, M.J., and Peng, K. (2010), "Cultural Differences in Expectations of Change and Tolerance for Contradiction: A Decade of Empirical Research," *Personality and Social Psychology Review*, 14(3), 296–312.

58 Reason, J.T., Manstead, A.S.R., and Stradling, S.G. (1990), "Errors and Violation on the Roads: A Real Distinction?" *Ergonomics*, 33(10–11), 1315–32.

59 Heine, S.J., and Hamamura, T. (2007), "In Search of East Asian Self-Enhancement," *Personality and Social Psychology Review*, 11(1), 4–27.

60 Santos, H.C., Varnum, M.E., and Grossmann, I. (2017), "Global Increases in Individualism," *Psychological Science*, 28(9), 1228–39. See also https://www.psychologicalscience.org/news/releases/individualistic-practices-and-values-increasing-around-the-world.html.

61 Franklin, B. (4 November 1789), Letter to John Wright. Unpublished, retrieved from http://franklinpapers.org/franklin/framedVolumes.jsp.

62 Franklin, B. (9 March 1790), To Ezra Stiles, with a statement of his religious creed. Retrieved from http://www.bartleby.com/400/prose/366.html.

Chapter 5

1 Kroc, R., with Anderson, R. (1977/87), *Grinding It Out: The Making of McDonald's*, New York: St Martin's Paperbacks, pp. 5–12, 39–59.

2 Quoted in Hastie, R., and Dawes, R.M. (2010), *Rational Choice in an Uncertain World: The Psychology of Judgment and Decision Making*, Thousand Oaks, CA: Sage, p. 66.

3 Damasio, A. (1994), *Descartes' Error*, New York: Avon Books, pp. 37–44.

4 This was also true when the participants viewed emotionally charged photographs. Unlike most participants, the people with damage to the frontal lobes showed no change to their skin conductance: Damasio, *Descartes' Error*, pp. 205–23.

5 Kandasamy, N., Garfinkel, S.N., Page, L., Hardy, B., Critchley, H.D., Gurnell, M., and Coates, J.M. (2016), "Interoceptive Ability Predicts Survival on a London Trading Floor," *Scientific Reports*, 6, 32986.

6 Werner, N.S., Jung, K., Duschek, S., and Schandry, R. (2009), "Enhanced Cardiac Perception Is Associated with Benefits in Decision-Making," *Psychophysiology*, 46(6), 1123–9.

7 Kandasamy et al., "Interoceptive Ability Predicts Survival on a London Trading Floor."

8 Ernst, J., Northoff, G., Böker, H., Seifritz, E., and Grimm, S. (2013), "Interoceptive Awareness Enhances Neural Activity during Empathy," *Human Brain Mapping*, 34(7), 1615–24. Terasawa, Y., Moriguchi, Y., Tochizawa, S., and Umeda, S. (2014), "Interoceptive Sensitivity Predicts Sensitivity to the Emotions of Others," *Cognition and Emotion*, 28(8), 1435–48.

9 Chua, E.F., and Bliss-Moreau, E. (2016), "Knowing Your Heart and Your Mind: The Relationships between Metamemory and Interoception," *Consciousness and Cognition*, 45, 146–58.

10 Umeda, S., Tochizawa, S., Shibata, M., and Terasawa, Y. (2016), "Prospective Memory Mediated by Interoceptive Accuracy: A Psychophysiological Approach," *Philosophical Transactions of the Royal Society B*, 371(1708), 20160005.

11 Kroc, with Anderson, *Grinding It Out*, p. 72. See also *Schupack v. McDonald's System, Inc.*, in which Kroc is quoted about his funny bone feeling: https://law.justia.com/cases/nebraska/supreme-court/1978/41114-1.html.

12 Hayashi, A.M. (2001), "When to Trust Your Gut," *Harvard Business Review*, 79(2), 59–65. See also Eugene Sadler-Smith's fascinating discussions of creativity and intuition in Sadler-Smith, E. (2010). *The Intuitive Mind: Profiting from the Power of Your Sixth Sense*. Chichester: John Wiley & Sons.

13 Feldman Barrett relates this story in her fascinating and engaging book, *How Emotions Are Made*: Feldman Barrett, L. (2017), *How Emotions Are Made*, London: Pan Macmillan, pp. 30–1. I've also written about this work for BBC Future: http://www.bbc.com/future/story/20171012-how-emotions-can-trick-your-mind-and-body.

14 Redelmeier, D.A., and Baxter, S.D. (2009), "Rainy Weather and Medical School Admission Interviews," *Canadian Medical Association Journal*, 181(12), 933.

15 Schnall, S., Haidt, J., Clore, G.L., and Jordan, A.H. (2008), "Disgust as Embodied Moral Judgment," *Personality and Social Psychology Bulletin*, 34(8), 1096–109.

16 Lerner, J.S., Li, Y., Valdesolo, P., and Kassam, K.S. (2015), "Emotion and Decision Making," *Annual Review of Psychology*, 66.

17 This quote was taken from Lisa Feldman Barrett's TED talk in Cambridge, 2018, https://www.youtube.com/watch?v=ZYAEh3T5a80.

18 Seo, M.G., and Barrett, L.F. (2007), "Being Emotional during Decision Making—Good or Bad? An Empirical Investigation," *Academy of Management Journal*, 50(4), 923–40.

19 Cameron, C.D., Payne, B.K., and Doris, J.M. (2013), "Morality in High Definition: Emotion Differentiation Calibrates the Influence of Incidental Disgust on Moral Judgments," *Journal of Experimental Social Psychology*, 49(4), 719–25. See also Fenton-O'Creevy, M., Soane, E., Nicholson, N., and Willman, P. (2011), "Thinking, Feeling and Deciding: The Influence of Emotions on the Decision Making and Performance of Traders," *Journal of Organizational Behavior*, 32(8), 1044–61.

20 See, for instance, Füstös, J., Gramann, K., Herbert, B.M., and Pollatos, O. (2012), "On the Embodiment of Emotion Regulation: Interoceptive Awareness Facilitates Reappraisal," *Social Cognitive and Affective Neuroscience*, 8(8), 911–17. And Kashdan, T.B., Barrett, L.F., and McKnight, P.E. (2015), "Unpacking Emotion Differentiation: Transforming Unpleasant Experience by Perceiving Distinctions in Negativity," *Current Directions in Psychological Science*, 24(1), 10–16.

21 See also Alkozei, A., Smith, R., Demers, L.A., Weber, M., Berryhill, S.M., and Killgore, W.D. (2018), "Increases in Emotional Intelligence after an Online Training Program Are Associated with Better Decision-Making on the Iowa Gambling Task," *Psychological Reports*, 0033294118771705.

22 Bruine de Bruin, W., Strough, J., and Parker, A.M. (2014), "Getting Older Isn't All

That Bad: Better Decisions and Coping When Facing 'Sunk Costs,'" *Psychology and Aging*, 29(3), 642.

23 Miu, A.C., and Crişan, L.G. (2011), "Cognitive Reappraisal Reduces the Susceptibility to the Framing Effect in Economic Decision Making," *Personality and Individual Differences*, 51(4), 478–82.

24 Halperin, E., Porat, R., Tamir, M., and Gross, J.J. (2013), "Can Emotion Regulation Change Political Attitudes in Intractable Conflicts? From the Laboratory to the Field," *Psychological Science*, 24(1), 106–11.

25 Grossmann, I., and Oakes, H. (2017), "Wisdom of Yoda and Mr. Spock: The Role of Emotions and the Self." Available as a pre-print on the PsyArxiv service: https://psyarxiv.com/jy5em/.

26 See, for instance, Hill, C.L., and Updegraff, J.A. (2012), "Mindfulness and Its Relationship to Emotional Regulation," *Emotion*, 12(1), 81. Daubenmier, J., Sze, J., Kerr, C.E., Kemeny, M.E., and Mehling, W. (2013), "Follow Your Breath: Respiratory Interoceptive Accuracy in Experienced Meditators," *Psychophysiology*, 50(8), 777–89. Fischer, D., Messner, M., and Pollatos, O. (2017), "Improvement of Interoceptive Processes after an 8-Week Body Scan Intervention," *Frontiers in Human Neuroscience*, 11, 452. Farb, N.A., Segal, Z.V., and Anderson, A.K. (2012), "Mindfulness Meditation Training Alters Cortical Representations of Interoceptive Attention," *Social Cognitive and Affective Neuroscience*, 8(1), 15–26.

27 Hafenbrack, A.C., Kinias, Z., and Barsade, S.G. (2014), "Debiasing the Mind through Meditation: Mindfulness and the Sunk-Cost Bias," *Psychological Science*, 25(2), 369–76.

28 For an in-depth discussion of the benefits of mindfulness to decision making, see Karelaia, N., and Reb, J. (2014), "Improving Decision Making through Mindfulness," forthcoming in Reb, J. and Atkins, P. (eds), *Mindfulness in Organizations*. Cambridge: Cambridge University Press. Hafenbrack, A.C. (2017), "Mindfulness Meditation as an On-the-Spot Workplace Intervention," *Journal of Business Research*, 75, 118–29.

29 Lakey, C.E., Kernis, M.H., Heppner, W.L., and Lance, C.E. (2008), "Individual Differences in Authenticity and Mindfulness as Predictors of Verbal Defensiveness," *Journal of Research in Personality*, 42(1), 230–8.

30 Reitz, M., Chaskalson, M., Olivier, S., and Waller, L. (2016), *The Mindful Leader*, Hult Research. Retrieved from: https://mbsr.co.uk/userfiles/Publications/Mindful-Leader-Report-2016-updated.pdf.

31 Kirk, U., Downar, J., and Montague, P.R. (2011), "Interoception Drives Increased Rational Decision-Making in Meditators Playing the Ultimatum Game," *Frontiers in Neuroscience*, 5, 49.

32 Yurtsever, G. (2008), "Negotiators' Profit Predicted by Cognitive Reappraisal, Suppression of Emotions, Misrepresentation of Information, and Tolerance of Ambiguity," *Perceptual and Motor Skills*, 106(2), 590–608.

33 Schirmer-Mokwa, K.L., Fard, P.R., Zamorano, A.M., Finkel, S., Birbaumer, N., and Kleber, B.A. (2015), "Evidence for Enhanced Interoceptive Accuracy in Professional Musicians," *Frontiers in Behavioral Neuroscience*, 9, 349. Christensen, J.F., Gaigg, S.B., and Calvo-Merino, B. (2018), "I Can Feel My Heartbeat: Dancers Have Increased Interoceptive Accuracy," *Psychophysiology*, 55(4), e13008.

34 Cameron, C.D., Payne, B.K., and Doris, J.M. (2013), "Morality in High Definition: Emotion Differentiation Calibrates the Influence of Incidental Disgust on Moral Judgments," *Journal of Experimental Social Psychology*, 49(4), 719–25.

35 Kircanski, K., Lieberman, M.D., and Craske, M.G. (2012), "Feelings into Words: Contributions of Language to Exposure Therapy," *Psychological Science*, 23(10), 1086–91.

36 According to Merriam-Webster's dictionary, the first written use of the word appeared in 1992 in *The London Magazine*, but it only recently entered more common usage: https://www.merriam-webster.com/words-at-play/hangry-meaning.

37 Zadie Smith offers us another lesson in emotion differentiation in the following essay on joy—a "strange admixture of terror, pain, and delight"—and the reasons that it should not be confused with pleasure. Besides being an astonishing read, the essay is a perfect illustration of the ways that we can carefully analyze our feelings and their effects on us. Smith, Z. (2013), "Joy," *New York Review of Books*, 60(1), 4.

38 Di Stefano, G., Gino, F., Pisano, G.P., and Staats, B.R. (2016), "Making Experience Count: The Role of Reflection in Individual Learning." Retrieved online at: http://dx.doi.org/10.2139/ssrn.2414478.

39 Cited in Pavlenko, A. (2014), *The Bilingual Mind*, Cambridge: Cambridge University Press, p. 282.

40 Keysar, B., Hayakawa, S.L., and An, S.G. (2012), "The Foreign-language Effect: Thinking in a Foreign Tongue Reduces Decision Biases," *Psychological Science*, 23(6), 661–8.

41 Costa, A., Foucart, A., Arnon, I., Aparici, M. and Apesteguia, J. (2014), "'Piensa' Twice: On the Foreign Language Effect in Decision Making," *Cognition*, 130(2), 236–54. See also Gao, S., Zika, O., Rogers, R.D., and Thierry, G. (2015), "Second Language Feedback Abolishes the 'Hot Hand' Effect during Even-Probability Gambling," *Journal of Neuroscience*, 35(15), 5983–9.

42 Caldwell-Harris, C.L. (2015), "Emotionality Differences between a Native and Foreign Language: Implications for Everyday Life," *Current Directions in Psychological Science*, 24(3), 214–19.

43 You can read more about these benefits in this accessible article by Amy Thompson, a professor of applied linguistics at the University of South Florida. Thompson, A. (12 December 2016), "How Learning a New Language Improves Tolerance," *The Conversation*: https://theconversation.com/how-learning-a-new-language-improves-tolerance-68472.

44 Newman-Toker, D.E., and Pronovost, P.J. (2009), "Diagnostic Errors—The Next Frontier for Patient Safety," *Journal of the American Medical Association*, 301(10), 1060–2.

45 Andrade, J. (2010), "What Does Doodling Do?" *Applied Cognitive Psychology*, 24(1), 100–6.

46 For Silvia Mamede's review and interpretation of these (and similar) results, see Mamede, S., and Schmidt, H.G. (2017), "Reflection in Medical Diagnosis: A Literature Review," *Health Professions Education*, 3(1), 15–25.

47 Schmidt, H.G., Van Gog, T., Schuit, S.C., Van den Berge, K., Van Daele, P.L., Bueving, H., Van der Zee, T., Van der Broek, W.W., Van Saase, J.L., and Mamede, S. (2017), "Do Patients' Disruptive Behaviours Influence the Accuracy of a Doctor's Diagnosis? A Randomised Experiment," *BMJ Quality & Safety*, 26(1), 19–23.

48 Schmidt, H.G., Mamede, S., Van den Berge, K., Van Gog, T., Van Saase, J.L., and Rikers R.M. (2014), "Exposure to Media Information about a Disease Can Cause Doctors to Misdiagnose Similar-Looking Clinical Cases," *Academic Medicine*, 89(2), 285–91.

49 For a further discussion of this new understanding of expertise, and the need for doctors to slow their thinking, see Moulton, C.A., Regehr G., Mylopoulos M., and MacRae, H.M. (2007), "Slowing Down When You Should: A New Model of Expert Judgment," *Academic Medicine*, 82(10Suppl.), S109–16.

50 Casey, P., Burke, K., and Leben, S. (2013), *Minding the Court: Enhancing the Decision-Making Process*, American Judges Association. Retrieved online from http://aja.ncsc.dni.us/pdfs/Minding-the-Court.pdf.

51 The first four stages of this model are usually attributed to Noel Burch of Gordan Training International.

52 The idea of "reflective competence" was originally proposed by David Baume, an education researcher at the Open University in the UK, who had described how experts need to be able to analyze and articulate their methods if they are to pass them on to others. But the doctor Pat Croskerry also uses the term to describe the fifth stage of expertise, in which experts can finally recognize the sources of their own bias.

Chapter 6

1 "Bananas and Flesh-eating Disease," Snopes.com. Retrieved 19 October 2017, http://www.snopes.com/medical/disease/bananas.asp.

2 Forster, K. (7 January 2017), "Revealed: How Dangerous Fake News Conquered Facebook," *Independent*: https://www.independent.co.uk/life-style/health-and-families/health-news/fake-news-health-facebook-cruel-damaging-social-media-mike-adams-natural-health-ranger-conspiracy-a7498201.html.

3 Binding, L. (24 July 2018), "India Asks Whatsapp to Curb Fake News Following SpateofLynchings,"SkyNewsonline:https://news.sky.com/story/india-asks-whatsapp-to-curb-fake-news-following-spate-of-lynchings-11425849.

4 Dewey, J. (1910), *How We Think*, p. 101. Mineola, NY: Dover Publications.

5 Galliford, N., and Furnham, A. (2017), "Individual Difference Factors and Beliefs in Medical and Political Conspiracy Theories," *Scandinavian Journal of Psychology*, 58(5), 422–8.

6. See, for instance, Kitai, E., Vinker, S., Sandiuk, A., Hornik, O., Zeltcer, C., and Gaver, A. (1998), "Use of Complementary and Alternative Medicine among Primary Care Patients," *Family Practice*, 15(5), 411–14. Molassiotis, A., et al. (2005), "Use of Complementary and Alternative Medicine in Cancer Patients: A European Survey," *Annals of Oncology*, 16(4), 655–63.
7 "Yes, We Have No Infected Bananas," CBC News, 6 March 2000: http://www.cbc.ca/news/canada/yes-we-have-no-infected-bananas-1.230298.
8 Rabin, N. (2006), "Interview with Stephen Colbert. *The Onion*, https://tv.avclub.com/stephen-colbert-1798208958.
9 Song, H., and Schwarz, N. (2008), "Fluency and the Detection of Misleading Questions: Low Processing Fluency Attenuates the Moses Illusion," *Social Cognition*, 26(6), 791–9.
10 This research is summarized in the following review articles: Schwarz, N., and Newman, E.J. (2017), "How Does the Gut Know Truth? The Psychology of 'Truthiness,'" *APA Science Brief*: http://www.apa.org/science/about/psa/2017/08/gut-truth.aspx. Schwarz, N., Newman, E., and Leach, W. (2016), "Making the Truth Stick & the Myths Fade: Lessons from Cognitive Psychology," *Behavioral Science & Policy*, 2(1), 85–95. See also Silva, R.R., Chrobot, N., Newman, E., Schwarz, N. and Topolinski, S. (2017), "Make It Short and Easy: Username Complexity Determines Trustworthiness Above and Beyond Objective Reputation," *Frontiers in Psychology*, 8, 2200.
11 Wu, W., Moreno, A.M., Tangen, J.M., and Reinhard, J. (2013), "Honeybees can discriminate between Monet and Picasso paintings," *Journal of Comparative Physiology A*, *199*(1), 45–55. Carlström, M., and Larsson, S.C. (2018). "Coffee consumption and reduced risk of developing type 2 diabetes: a systematic review with meta-analysis," *Nutrition Reviews*, 76(6), 395–417. Olszewski, M., and Ortolano, R. (2011). "Knuckle cracking and hand osteoarthritis," *The Journal of the American Board of Family Medicine*, 24(2), 169–174.
12 Newman, E.J., Garry, M., and Bernstein, D.M., et al. (2012), "Nonprobative Words (or Photographs) Inflate Truthiness," *Psychonomic Bulletin and Review*, 19(5), 969–74.
13 Weaver, K., Garcia, S.M., Schwarz, N., and Miller, D.T. (2007), "Inferring the Popularity of an Opinion from Its Familiarity: A Repetitive Voice Can Sound Like a Chorus," *Journal of Personality and Social Psychology*, 92(5), 821–33.
14 Weisbuch, M., and Mackie, D. (2009), "False Fame, Perceptual Clarity, or Persuasion? Flexible Fluency Attribution in Spokesperson Familiarity Effects," *Journal of Consumer Psychology*, 19(1), 62–72.
15 Fernandez-Duque, D., Evans, J., Christian, C., and Hodges, S.D. (2015), "Superfluous Neuroscience Information Makes Explanations of Psychological Phenomena More Appealing," *Journal of Cognitive Neuroscience*, 27(5), 926–44.
16 Proctor, R. (2011), *Golden Holocaust: Origins of the Cigarette Catastrophe and the Case for Abolition*. Berkeley: University of California Press, p. 292.
17 See the following paper for a thorough discussion of this effect: Schwarz, N., Sanna, L.J., Skurnik, I., and Yoon, C. (2007), "Metacognitive Experiences and the

Intricacies of Setting People Straight: Implications for Debiasing and Public Information Campaigns," *Advances in Experimental Social Psychology*, 39, 127–61. See also Pluviano, S., Watt, C., and Della Sala, S. (2017), "Misinformation Lingers in Memory: Failure of Three Pro-Vaccination Strategies," *PLOS One*, 12(7), e0181640.

18 Glum, J. (11 November 2017), "Some Republicans Still Think Obama Was Born in Kenya as Trump Resurrects Birther Conspiracy Theory," *Newsweek*, http://www.newsweek.com/trump-birther-obama-poll-republicans-kenya-744195.

19 Lewandowsky, S., Ecker, U. K., Seifert, C. M., Schwarz, N., and Cook, J. (2012). "Misinformation and its Correction: Continued Influence and Successful Debiasing." *Psychological Science in the Public Interest*, 13(3), 106–131.

20 Cook, J., and Lewandowsky, S. (2011), *The Debunking Handbook*. Available at https://skepticalscience.com/docs/Debunking_Handbook.pdf.

21 NHS Choices, "10 Myths about the Flu and Flu Vaccine," https://www.nhs.uk/Livewell/winterhealth/Pages/Flu-myths.aspx.

22 Smith, I.M., and MacDonald, N.E. (2017), "Countering Evidence Denial and the Promotion of Pseudoscience in Autism Spectrum Disorder," *Autism Research*, 10(8), 1334–7.

23 Pennycook, G., Cheyne, J.A., Koehler, D.J., et al. (2016), "Is the Cognitive Reflection Test a Measure of Both Reflection and Intuition?" *Behavior Research Methods*, 48(1), 341–8.

24 Pennycook, G. (2014), "Evidence That Analytic Cognitive Style Influences Religious Belief: Comment on Razmyar and Reeve (2013)," *Intelligence*, 43, 21–6.

25 Much of this work on the Cognitive Reflection Test has been summarized in the following review paper: Pennycook, G., Fugelsang, J.A., and Koehler, D.J. (2015), "Everyday Consequences of Analytic Thinking," *Current Directions in Psychological Science*, 24(6), 425–32.

26 Pennycook, G., Cheyne, J.A., Barr, N., Koehler, D.J., and Fugelsang, J.A. (2015), "On the Reception and Detection of Pseudo-Profound Bullshit," *Judgment and Decision Making*, 10(6), 549–63.

27 Pennycook, G., and Rand, D.G. (2018), "Lazy, Not Biased: Susceptibility to Partisan Fake News Is Better Explained by Lack of Reasoning than by Motivated Reasoning," *Cognition*, https://doi.org/10.1016/j.cognition.2018.06.011. See also Pennycook, G., and Rand, D.G. (2017), "Who Falls for Fake News? The Roles of Bullshit Receptivity, Overclaiming, Familiarity, and Analytic Thinking," unpublished paper, https://dx.doi.org/10.2139/ssrn.3023545.

28 Swami, V., Voracek, M., Stieger, S., Tran, U.S., and Furnham, A. (2014), "Analytic Thinking Reduces Belief in Conspiracy Theories," *Cognition*, 133(3), 572–85. This method of priming analytic thought can also reduce religious beliefs and paranormal thinking: Gervais, W.M., and Norenzayan, A. (2012), "Analytic Thinking Promotes Religious Disbelief," *Science*, 336(6080), 493–6.

29 Long before the modern form of the Cognitive Reflection Test was invented, psychologists had noticed that it may be possible to prime people to be more critical of the information they receive. In 1987, participants were given deceptively simple trivia questions, with a tempting misleading answer. The process cured their overconfidence on a subsequent test, helping them to calibrate their confidence to their actual knowledge. Arkes, H.R., Christensen, C., Lai, C., and Blumer, C. (1987), "Two Methods of Reducing Overconfidence," *Organizational Behavior and Human Decision Processes*, 39(1), 133–44.

30 Fitzgerald, C.J., and Lueke, A.K. (2017), "Mindfulness Increases Analytical Thought and Decreases Just World Beliefs," *Current Research in Social Psychology*, 24(8), 80–5.

31 Robinson himself admitted that the names were unverified. Hebert, H.J. (1 May 1998), "Odd Names Added to Greenhouse Plea," Associated Press: https://apnews.com/aec8beea85d7fe76fc9cc77b8392d79e.

32 Cook, J., Lewandowsky, S., and Ecker, U.K. (2017), "Neutralizing Misinformation through Inoculation: Exposing Misleading Argumentation Techniques Reduces Their Influence," *PLOS One*, 12(5), e0175799.

33 For further evidence, see Roozenbeek, J., and Van der Linden, S. (2018), "The Fake News Game: Actively Inoculating against the Risk of Misinformation," *Journal of Risk Research*. DOI: 10.1080/13669877.2018.1443491.

34 McLaughlin, A.C., and McGill, A.E. (2017), "Explicitly Teaching Critical Thinking Skills in a History Course," *Science and Education*, 26(1–2), 93–105. For a further discussion of the benefits of inoculation in education, see Schmaltz, R., and Lilienfeld, S.O. (2014), "Hauntings, Homeopathy, and the Hopkinsville Goblins: Using Pseudoscience to Teach Scientific Thinking," *Frontiers in Psychology*, 5, 336.

35 Rowe, M.P., Gillespie, B.M., Harris, K.R., Koether, S.D., Shannon, L.J.Y., and Rose, L.A. (2015), "Redesigning a General Education Science Course to Promote Critical Thinking," *CBE-Life Sciences Education*, 14(3), ar30.

36 See, for instance, Butler, H.A. (2012), "Halpern Critical Thinking Assessment Predicts Real-World Outcomes of Critical Thinking," *Applied Cognitive Psychology*, 26(5), 721–9. Butler, H.A., Pentoney, C., and Bong, M.P. (2017), "Predicting Real-World Outcomes: Critical Thinking Ability Is a Better Predictor of Life Decisions than Intelligence," *Thinking Skills and Creativity*, 25, 38–46.

37 See, for instance, Arum, R., and Roksa, J. (2011), *Academically Adrift: Limited Learning on College Campuses*, Chicago, IL: University of Chicago Press.

38 "Editorial: Louisiana's Latest Assault on Darwin," *New York Times* (21 June 2008), https://www.nytimes.com/2008/06/21/opinion/21sat4.html.

39 Kahan, D.M. (2016), "The Politically Motivated Reasoning Paradigm, Part 1: What Politically Motivated Reasoning Is and How to Measure It," in *Emerging Trends in the Social and Behavioral Sciences: An Interdisciplinary, Searchable, and Linkable Resource*, doi: 10.1002/9781118900772.

40 Hope, C. (8 June 2015), "Campaigning against GM Crops Is 'Morally Unacceptable,' Says Former Greenpeace Chief," *Daily Telegraph*, http://www.telegraph.co

.uk/news/earth/agriculture/crops/11661016/Campaigning-against-GM-crops-is-morally-unacceptable-says-former-Greenpeace-chief.html.

41 Shermer, M. (2007), *Why People Believe Weird Things*, London: Souvenir Press, pp. 13–15. (Originally published 1997.)

42 You can read one of these exchanges on the *Skeptic* website: https://www.skeptic.com/eskeptic/05-05-03/.

43 You can find the Skepticism 101 syllabus, including a reading list, at the following website: https://www.skeptic.com/downloads/Skepticism101-How-to-Think-Like-a-Scientist.pdf.

44 For more information, see Shermer, M. (2012), *The Believing Brain*, London: Robinson, pp. 251–8. If you are looking for a course in inoculation, I'd thoroughly recommend reading Shermer's work.

Chapter 7

1 Feynman, R. (1985), *Surely You're Joking, Mr. Feynman!: Adventures of a Curious Character*, New York: W. W. Norton.

2 The following article, an interview with one of Feynman's former students, offers this interpretation: Wai, J. (2011), "A Polymath Physicist on Richard Feynman's 'Low' IQ and Finding another Einstein," *Psychology Today*, https://www.psychologytoday.com/blog/finding-the-next-einstein/201112/polymath-physicist-richard-feynmans-low-iq-and-finding-another.

3 Gleick, J. (1992), *Genius: Richard Feynman and Modern Physics* (Kindle Edition), pp. 30–5.

4 Gleick, J. (17 February 1988), "Richard Feynman Dead at 69: Leading Theoretical Physicist," *New York Times*, http://www.nytimes.com/1988/02/17/obituaries/richard-feynman-dead-at-69-leading-theoretical-physicist.html?pagewanted=all.

5 The Nobel Prize in Physics 1965: https://www.nobelprize.org/nobel_prizes/physics/laureates/1965/.

6 Kac, M. (1987), *Enigmas of Chance: An Autobiography*, Berkeley: University of California Press, p. xxv.

7 Gleick, "Richard Feynman Dead at 69."

8 Feynman, R.P. (1999), *The Pleasure of Finding Things Out*, New York: Perseus Books, p. 3.

9 Feynman, R.P. (2006), *Don't You Have Time to Think*, ed. Feynman, M., London: Penguin, p. 414.

10 Darwin, C. (2016), *Life and Letters of Charles Darwin* (Vol. 1), Krill Press via PublishDrive. Available online at: https://charles-darwin.classic-literature.co.uk/the-life-and-letters-of-charles-darwin-volume-i/ebook-page-42.asp.

11 Darwin, C. (1958), *Selected Letters on Evolution and Natural Selection*, ed. Darwin, F., New York: Dover Publications, p. 9.

12 Engel, S. (2011), "Children's Need to Know: Curiosity in Schools," *Harvard Educational Review*, 81(4), 625–45.

13 Engel, S. (2015), *The Hungry Mind*, Cambridge, MA: Harvard University Press, p. 3.

14 Von Stumm, S., Hell, B., and Chamorro-Premuzic, T. (2011), "The Hungry Mind: Intellectual Curiosity Is the Third Pillar of Academic Performance," *Perspectives on Psychological Science*, 6(6), 574–88.

15 Engel, "Children's Need to Know."

16 Gruber, M.J., Gelman, B.D., and Ranganath, C. (2014), "States of Curiosity Modulate Hippocampus-Dependent Learning via the Dopaminergic Circuit," *Neuron*, 84(2), 486–96. A previous, slightly less detailed study had come to broadly the same conclusions: Kang, M.J., Hsu, M., Krajbich, I.M., Loewenstein, G., McClure, S.M., Wang, J.T.Y., and Camerer, C.F. (2009), "The Wick in the Candle of Learning: Epistemic Curiosity Activates Reward Circuitry and Enhances Memory," *Psychological Science*, 20(8), 963–73.

17 Hardy III, J.H., Ness, A.M., and Mecca, J. (2017), "Outside the Box: Epistemic Curiosity as a Predictor of Creative Problem Solving and Creative Performance," *Personality and Individual Differences*, 104, 230–7.

18 Leonard, N.H., and Harvey, M. (2007), "The Trait of Curiosity as a Predictor of Emotional Intelligence," *Journal of Applied Social Psychology*, 37(8), 1914–29.

19 Sheldon, K.M., Jose, P.E., Kashdan, T.B., and Jarden, A. (2015), "Personality, Effective Goal-Striving, and Enhanced Well-Being: Comparing 10 Candidate Personality Strengths," *Personality and Social Psychology Bulletin*, 41(4), 575–85.

20 Kashdan, T.B., Gallagher, M.W., Silvia, P.J., Winterstein, B.P., Breen, W.E., Terhar, D., and Steger, M.F. (2009), "The Curiosity and Exploration Inventory-II: Development, Factor Structure, and Psychometrics." *Journal of Research in Personality*, 43(6), 987–98.

21 Krakovsky, M. (2007), "The Effort Effect," *Stanford Alumni* magazine, https://alumni.stanford.edu/get/page/magazine/article/?article_id=32124.

22 Trei, L. (2007), "New Study Yields Instructive Results on How Mindset Affects Learning," Stanford News website, https://news.stanford.edu/news/2007/february7/dweck-020707.html.

23 Harvard Business Review staff (2014), "How Companies Can Profit from a 'Growth Mindset,'" *Harvard Business Review*, https://hbr.org/2014/11/how-companies-can-profit-from-a-growth-mindset.

24 Along these lines, a recent study found that gifted students are particularly at risk of the fixed mind-set: Esparza, J., Shumow, L., and Schmidt, J.A. (2014), "Growth Mindset of Gifted Seventh Grade Students in Science," *NCSSSMST Journal*, 19(1), 6–13.

25 Dweck, C. (2012), *Mindset: Changing the Way You Think to Fulfil Your Potential*, London: Robinson, pp. 17–18, 234–9.

26 Mangels, J.A., Butterfield, B., Lamb, J., Good, C., and Dweck, C.S. (2006), "Why Do Beliefs about Intelligence Influence Learning Success? A Social Cognitive Neuroscience Model," *Social Cognitive and Affective Neuroscience*, 1(2), 75–86.

27 Claro, S., Paunesku, D., and Dweck, C.S. (2016), "Growth Mindset Tempers the

Effects of Poverty on Academic Achievement," *Proceedings of the National Academy of Sciences*, 113(31), 8664–8.

28 For evidence of the benefits of mind-set, see the following meta-analysis, examining 113 studies: Burnette, J.L., O'Boyle, E.H., VanEpps, E.M., Pollack, J.M., and Finkel, E.J. (2013), "Mind-sets Matter: A Meta-Analytic Review of Implicit Theories and Self-regulation," *Psychological Bulletin*, 139(3), 655–701.

29 Quoted in Roberts, R., and Kreuz, R. (2015), *Becoming Fluent: How Cognitive Science Can Help Adults Learn a Foreign Language*, Cambridge, MA: MIT Press, pp. 26–7.

30 See, for example, Rustin, S. (10 May 2016), "New Test for 'Growth Mindset,' the Theory That Anyone Who Tries Can Succeed," *Guardian*, https://www.theguardian.com/education/2016/may/10/growth-mindset-research-uk-schools-sats.

31 Brummelman, E., Thomaes, S., Orobio de Castro, B., Overbeek, G., and Bushman, B.J. (2014), "'That's Not Just Beautiful—That's Incredibly Beautiful!' The Adverse Impact of Inflated Praise on Children with Low Self-Esteem," *Psychological Science*, 25(3), 728–35.

32 Dweck, C. (2012), *Mindset*, London: Robinson, pp. 180–6, 234–9. See also Haimovitz, K. and Dweck, C.S. (2017), "The Origins of Children's Growth and Fixed Mindsets: New Research and a New Proposal," *Child Development*, 88(6), 1849–59.

33 Frank, R. (16 October 2013), "Billionare Sara Blakely Says Secret to Success Is Failure,"CNBC:https://www.cnbc.com/2013/10/16/billionaire-sara-blakely-says-secret-to-success-is-failure.html.

34 See, for instance, Paunesku, D., Walton, G.M., Romero, C., Smith, E.N., Yeager, D.S., and Dweck, C.S. (2015), "Mind-Set Interventions Are a Scalable Treatment for Academic Underachievement," *Psychological Science*, 26(6), 784–93. For further evidence of the power of interventions, see the following meta-analysis: Lazowski, R.A., and Hulleman, C.S. (2016), "Motivation Interventions in Education: A Meta-Analytic Review," *Review of Educational Research*, 86(2), 602–40.

35 See, for instance, the following meta-analysis, which found a small but significant effect: Sisk, V.F., Burgoyne, A.P., Sun, J., Butler, J.L., and Macnamara, B.N. (2018), "To What Extent and Under Which Circumstances Are Growth Mind-Sets Important to Academic Achievement? Two Meta-Analyses," *Psychological Science*, in press, https://doi.org/10.1177/0956797617739704. The exact interpretation of these results is still the matter of debate. Generally speaking, it would seem that the growth mind-set is most important when students feel vulnerable/threatened—meaning that the interventions are most effective for children of poorer households, for instance. And while one-shot interventions do provide some long-term benefits, it seems clear that a more regular program would be needed to maintain larger effects. See: Orosz, G., Péter-Szarka, S., Bőthe, B., Tóth-Király, I., and Berger, R. (2017), "How Not to Do a Mindset Intervention: Learning From a Mindset Intervention among Students with Good Grades," *Frontiers in Psychology*, 8, 311. For an independent analysis, see the following blog on the British Psychological

Society website: https://digest.bps.org.uk/2018/03/23/this-cheap-brief-growth-mindset-intervention-shifted-struggling-students-onto-a-more-successful-trajectory/.

36 Feynman, R.P., and Feynman, Michelle (2006), *Don't You Have Time to Think?*, London: Penguin.

37 This episode is described in greater detail in Feynman's memoir, *Surely You're Joking, Mr. Feynman!*

38 Feynman and Feynman, *Don't You Have Time to Think?*, p. xxi.

39 Feynman, R. (1972), *Nobel Lectures, Physics 1963–1970*, Amsterdam: Elsevier. Retrieved online at https://www.nobelprize.org/nobel_prizes/physics/laureates/1965/feynman-lecture.html.

40 Kahan, D.M., Landrum, A., Carpenter, K., Helft, L., and Hall Jamieson, K. (2017), "Science Curiosity and Political Information Processing," *Political Psychology*, 38(S1), 179–99.

41 Kahan, D. (2016), "Science Curiosity and Identity-Protective Cognition . . . A Glimpse at a Possible (Negative) Relationship," Cultural Cognition Project blog, http://www.culturalcognition.net/blog/2016/2/25/science-curiosity-and-identity-protective-cognition-a-glimps.html.

42 Porter, T., and Schumann, K. (2017), "Intellectual Humility and Openness to the Opposing View," *Self and Identity*, 17(2), 1–24. Igor Grossmann, incidentally, has come to similar conclusions in one of his most recent studies: Brienza, J.P., Kung, F.Y.H., Santos, H.C., Bobocel, D.R., and Grossmann, I. (2017), "Wisdom, Bias, and Balance: Toward a Process-Sensitive Measurement of Wisdom-Related Cognition." *Journal of Personality and Social Psychology*, advance online publication, http://dx.doi.org/10.1037/pspp0000171.

43 Brienza, Kung, Santos, Bobocel and Grossmann, "Wisdom, Bias, and Balance."

44 Feynman, R.P. (2015), *The Quotable Feynman*, Princeton, NJ: Princeton University Press, p. 283.

45 Morgan, E.S. (2003), *Benjamin Franklin*, New Haven, CT: Yale University Press, p. 6.

46 Friend, T. (13 November 2017), "Getting On," *New Yorker*, https://www.newyorker.com/magazine/2017/11/20/why-ageism-never-gets-old/amp.

47 Friedman, T.L. (22 February 2014), "How to Get a Job at Google," *New York Times*, https://www.nytimes.com/2014/02/23/opinion/sunday/friedman-how-to-get-a-job-at-google.html.

Chapter 8

1 Details of this story can be found in one of Stigler's earliest works: Stevenson, H.W., and Stigler, J.W. (1992), *The Learning Gap*, New York: Summit Books, p. 16.

2 Waldow, F., Takayama, K. and Sung, Y.K. (2014), "Rethinking the Pattern of External Policy Referencing: Media Discourses over the 'Asian Tigers'" PISA Success in Australia, Germany and South Korea," *Comparative Education*, 50(3), 302–21.

3 Baddeley, A.D., and Longman, D.J.A. (1978), "The Influence of Length and Fre-

quency of Training Session on the Rate of Learning to Type," *Ergonomics*, 21(8), 627–35.

4 Rohrer, D. (2012), "Interleaving Helps Students Distinguish among Similar Concepts," *Educational Psychology Review*, 24(3), 355–67.

5 See, for instance, Kornell, N., Hays, M.J., and Bjork, R.A. (2009), "Unsuccessful Retrieval Attempts Enhance Subsequent Learning," *Journal of Experimental Psychology: Learning, Memory, and Cognition*, 35(4), 989. DeCaro, M.S. (2018), "Reverse the Routine: Problem Solving before Instruction Improves Conceptual Knowledge in Undergraduate Physics," *Contemporary Educational Psychology*, 52, 36–47. Clark, C.M. and Bjork, R.A. (2014), "When and Why Introducing Difficulties and Errors Can Enhance Instruction," in Benassi, V.A., Overson, C.E., and Hakala, C.M. (eds), *Applying Science of Learning in Education: Infusing Psychological Science into the Curriculum*, Washington, DC: Society for the Teaching of Psychology, pp. 20–30.

6 See, for instance, Kapur, Manu (2010), "Productive Failure in Mathematical Problem Solving," *Instructional Science*, 38(6), 523–50. And Overoye, A.L., and Storm, B.C. (2015), "Harnessing the Power of Uncertainty to Enhance Learning," *Translational Issues in Psychological Science*, 1(2), 140.

7 See, for instance, Susan Engel's discussion of Ruth Graner and Rachel Brown's work in Engel, S. (2015), *The Hungry Mind*, Cambridge, MA: Harvard University Press, p. 118.

8 See the following articles for a review of these metacognitive illusions: Bjork, R.A., Dunlosky, J., and Kornell, N. (2013), "Self-Regulated Learning: Beliefs, Techniques, and Illusions," *Annual Review of Psychology*, 64, 417–44. Yan, V.X., Bjork, E.L. and Bjork, R.A. (2016), "On the Difficulty of Mending Metacognitive Illusions: A Priori Theories, Fluency Effects, and Misattributions of the Interleaving Benefit," *Journal of Experimental Psychology: General*, 145(7), 918–33.

9 For more information about these results, read Stigler's earlier and later research. Hiebert, J., and Stigler, J.W. (2017), "The Culture of Teaching: A Global Perspective," in *International Handbook of Teacher Quality and Policy*, Abingdon, England: Routledge, pp. 62–75. Stigler, J.W., and Hiebert, J. (2009), *The Teaching Gap: Best Ideas from the World's Teachers for Improving Education in the Classroom*, New York: Simon & Schuster.

10 Park, H. (2010), "Japanese and Korean High Schools and Students in Comparative Perspective," in *Quality and Inequality of Education*, Netherlands: Springer, pp. 255–73.

11 Hiebert and Stigler, "The Culture of Teaching."

12 For a more in-depth discussion of these ideas, see Byrnes, J.P., and Dunbar, K.N. (2014), "The Nature and Development of Critical-Analytic Thinking," *Educational Psychology Review*, 26(4), 477–93.

13 See the following for further cross-cultural comparisons: Davies, M., and Barnett, R. (eds) (2015), *The Palgrave Handbook of Critical Thinking in Higher Education*, Netherlands: Springer.

14 For a comprehensive summary of this work, see Spencer-Rodgers, J., Williams, M.J., and Peng, K. (2010), "Cultural Differences in Expectations of Change and Tolerance for Contradiction: A Decade of Empirical Research," *Personality and Social Psychology Review*, 14(3), 296–312.

15 Rowe, M.B. (1986), "Wait Time: Slowing Down May Be a Way of Speeding Up!" *Journal of Teacher Education*, 37(1), 43–50.

16 Langer, E. (1997), *The Power of Mindful Learning*, Reading, MA: Addison-Wesley, p. 18.

17 Ritchhart, R., and Perkins, D.N. (2000), "Life in the Mindful Classroom: Nurturing the Disposition of Mindfulness," *Journal of Social Issues*, 56(1), 27-47.

18 Although Langer has pioneered the benefits of ambiguity in education, other scientists have shown comparably powerful results. Similarly, Robert S. Siegler and Xiaodong Lin have found that children learn mathematics and physics best when they are asked to think through *right* and *wrong* answers to problems, since this encourages them to consider alternative strategies and to identify ineffective ways of thinking. See, for example, Siegler, R.S., and Lin, X., "Self-Explanations Promote Children's Learning," in Borkowski, J.G., Waters, H.S. and Schneider, W. (eds) (2010), *Metacognition, Strategy Use, and Instruction*, New York: Guilford, pp. 86-113.

19 Langer, *The Power of Mindful Learning*, p. 29. And Overoye, A.L., and Storm, B.C. (2015), "Harnessing the Power of Uncertainty to Enhance Learning," *Translational Issues in Psychological Science*, 1(2), 140. See also Engel, S. (2011), "Children's Need to Know: Curiosity in Schools," *Harvard Educational Review*, 81(4), 625–45.

20 Brackett, M.A., Rivers, S.E., Reyes, M.R., and Salovey, P. (2012), "Enhancing Academic Performance and Social and Emotional Competence with the RULER Feeling Words Curriculum," *Learning and Individual Differences*, 22(2), 218–24. Jacobson, D., Parker, A., Spetzler, C., De Bruin, W.B., Hollenbeck, K., Heckerman, D., and Fischhoff, B. (2012), "Improved Learning in US History and Decision Competence with Decision-focused Curriculum," *PLOS One*, 7(9), e45775.

21 The following study, for instance, shows that greater intellectual humility completely eradicates the differences in academic achievement that you normally find between high- and low-ability students. Hu, J., Erdogan, B., Jiang, K., Bauer, T.N., and Liu, S. (2018), "Leader Humility and Team Creativity: The Role of Team Information Sharing, Psychological Safety, and Power Distance," *Journal of Applied Psychology*, 103(3), 313.

22 For the sources of these suggestions, see Bjork, Dunlosky, and Kornell, "Self-regulated Learning." Soderstrom, N.C., and Bjork, R.A. (2015), "Learning versus Performance: An Integrative Review," *Perspectives on Psychological Science*, 10(2), 176–99. Benassi, V.A., Overson, C., and Hakala, C.M. (2014), *Applying Science of Learning in Education: Infusing Psychological Science into the Curriculum*, American Psychological Association.

23 See, for example, the following blog post by psychologist and musician Christine

Carter for more information: https://bulletproofmusician.com/why-the-progress-in-the-practice-room-seems-to-disappear-overnight/.

24 Langer, E., Russel, T., and Eisenkraft, N. (2009), "Orchestral Performance and the Footprint of Mindfulness," *Psychology of Music*, 37(2), 125–36.

25 These data have been released on the IVA's website: http://www.ivalongbeach.org/academics/curriculum/61-academics/test-scores-smarter-balanced.

Chapter 9

1 Taylor, D. (2016, June 27), "England Humiliated as Iceland Knock Them Out of Euro 2016," *Guardian*, https://www.theguardian.com/football/2016/jun/27/england-iceland-euro-2016-match-report.

2 See, for example, http://www.independent.co.uk/sport/football/international/england-vs-iceland-steve-mcclaren-reaction-goal-euro-2016-a7106896.html.

3 Taylor, "England Humiliated as Iceland Knock Them Out of Euro 2016."

4 Wall, K. (27 June 2016), "Iceland Wins Hearts at Euro 2016 as Soccer's Global Underdog," *Time*, http://time.com/4383403/iceland-soccer-euro-2016-england/.

5 Zeileis, A., Leitner, C., and Hornik, K. (2016), "Predictive Bookmaker Consensus Model for the UEFA Euro 2016," *Working Papers in Economics and Statistics*, No. 2016–15, https://www.econstor.eu/bitstream/10419/146132/1/859777529.pdf.

6 Woolley, A.W., Aggarwal, I., and Malone, T.W. (2015), "Collective Intelligence and Group Performance," *Current Directions in Psychological Science*, 24(6), 420–4.

7 Wuchty, S., Jones, B.F., and Uzzi, B. (2007), "The Increasing Dominance of Teams in Production of Knowledge," *Science*, 316(5827), 1036–9.

8 Woolley, A.W., Chabris, C.F., Pentland, A., Hashmi, N., and Malone, T.W. (2010), "Evidence for a Collective Intelligence Factor in the Performance of Human Groups," *Science*, 330(6004), 686–8.

9 Engel, D., Woolley, A.W., Jing, L.X., Chabris, C.F., and Malone, T.W. (2014), "Reading the Mind in the Eyes or Reading between the Lines? Theory of Mind Predicts Collective Intelligence Equally Well Online and Face-to-face," *PLOS One*, 9(12), e115212.

10 Mayo, A.T., and Woolley, A.W. (2016), "Teamwork in Health Care: Maximizing Collective Intelligence via Inclusive Collaboration and Open Communication," *AMA Journal of Ethics*, 18(9), 933–40.

11 Woolley, Aggarwal, and Malone (2015), "Collective Intelligence and Group Performance."

12 Kim, Y.J., Engel, D., Woolley, A.W., Lin, J.Y.T., McArthur, N., and Malone, T.W. (2017), "What Makes a Strong Team? Using Collective Intelligence to Predict Team Performance in League of Legends," in *Proceedings of the 2017 ACM Conference on Computer Supported Cooperative Work and Social Computing*, New York: ACM, pp. 2316–29.

13 See, for example, Ready, D.A., and Conger, J.A. (2007), "Make Your Company a Talent Factory," *Harvard Business Review*, 85(6), 68–77. The following paper also

offers original surveys as well as a broader discussion of our preference for "talent" at the expense of teamwork: Swaab, R.I., Schaerer, M., Anicich, E.M., Ronay, R., and Galinsky, A.D. (2014), "The Too-Much-Talent Effect: Team Interdependence Determines When More Talent Is Too Much or Not Enough," *Psychological Science*, 25(8), 1581–91. See also Alvesson, M., and Spicer, A. (2016), *The Stupidity Paradox: The Power and Pitfalls of Functional Stupidity at Work*, London: Profile, Kindle Edition (Kindle location 1492–1504).

14 This work was a collaboration with Cameron Anderson, also at the University of California, Berkeley. Hildreth, J.A.D. and Anderson, C. (2016), "Failure at the Top: How Power Undermines Collaborative Performance," *Journal of Personality and Social Psychology*, 110(2), 261–86.

15 Greer, L.L., Caruso, H.M., and Jehn, K.A. (2011), "The Bigger They Are, the Harder They Fall: Linking Team Power, Team Conflict, and Performance," *Organizational Behavior and Human Decision Processes*, 116(1), 116–28.

16 Groysberg, B., Polzer, J.T., and Elfenbein, H.A. (2011), "Too Many Cooks Spoil the Broth: How High-Status Individuals Decrease Group Effectiveness," *Organization Science*, 22(3), 722–37.

17 Kishida, K.T., Yang, D., Quartz, K.H., Quartz, S.R., and Montague, P.R. (2012), "Implicit Signals in Small Group Settings and Their Impact on the Expression of Cognitive Capacity and Associated Brain Responses," *Philosophical Transactions of the Royal Society B*, 367(1589), 704–16.

18 "Group Settings Can Diminish Expressions of Intelligence, Especially among Women," Virginia Tech Carilion Research Institute, http://research.vtc.vt.edu/news/2012/jan/22/group-settings-can-diminish-expressions-intelligen/.

19 Galinsky, A., and Schweitzer, M. (2015), "The Problem of Too Much Talent," *The Atlantic*, https://www.theatlantic.com/business/archive/2015/09/hierarchy-friend-foe-too-much-talent/401150/

20 Swaab, R.I., Schaerer, M., Anicich, E.M., Ronay, R., and Galinsky, A.D. (2014), "The Too-Much-Talent Effect."

21 Herbert, I. (27 June 2016), "England vs Iceland: Too Wealthy, Too Famous, Too Much Ego—Joe Hart Epitomises Everything That's Wrong," *Independent*, http://www.independent.co.uk/sport/football/international/england-vs-iceland-reaction-too-rich-too-famous-too-much-ego-joe-hart-epitomises-everything-that-is-a7106591.html.

22 Roberto, M.A. (2002), "Lessons from Everest: The Interaction of Cognitive Bias, Psychological Safety, and System Complexity," *California Management Review*, 45(1), 136–58.

23 https://www.pbs.org/wgbh/pages/frontline/everest/stories/leadership.html.

24 Schwartz, S. (2008), "The 7 Schwartz Cultural Value Orientation Scores for 80 Countries," doi: 10.13140/RG.2.1.3313.3040.

25 Anicich, E.M., Swaab, R.I., and Galinsky, A.D. (2015). Hierarchical cultural values predict success and mortality in high-stakes teams. *Proceedings of the National Academy of Sciences*, 112(5), 1338–43.

26 Jang, S. (2017). "Cultural Brokerage and Creative Performance in Multicultural Teams," *Organization Science*, 28(6), 993–1009.

27 Ou, A.Y., Waldman, D.A., and Peterson, S.J. (2015), "Do Humble CEOs Matter? An Examination of CEO Humility and Firm Outcomes," *Journal of Management*, 44(3), 1147–73. For further discussion of the benefits of humble leaders, and the reasons we don't often value their humility, see Mayo, M. (2017), "If Humble People Make the Best Leaders, Why Do We Fall for Charismatic Narcissists?" *Harvard Business Review*, https://hbr.org/2017/04/if-humble-people-make-the-best-leaders-why-do-we-fall-for-charismatic-narcissists. Heyden, M.L.M. and Hayward, M. (2017), "It's Hard to Find a Humble CEO: Here's Why," *The Conversation*, https://theconversation.com/its-hard-to-find-a-humble-ceo-heres-why-81951. Rego, A., Owens, B., Leal, S., Melo, A.I., e Cunha, M.P., Gonçalves, L., and Ribeiro, P. (2017), "How Leader Humility Helps Teams to Be Humbler, Psychologically Stronger, and More Effective: A Moderated Mediation Model," *The Leadership Quarterly*, 28(5), 639–58.

28 Cable, D. (23 April 2018), "How Humble Leadership Really Works," *Harvard Business Review*, https://hbr.org/2018/04/how-humble-leadership-really-works.

29 Rieke, M., Hammermeister, J. and Chase, M. (2008), "Servant Leadership in Sport: A New Paradigm for Effective Coach Behavior," *International Journal of Sports Science & Coaching*, 3(2), 227–39.

30 Abdul-Jabbar, K. (2017), *Coach Wooden and Me: Our 50-Year Friendship On and Off the Court*, New York: Grand Central. The following article provides a more in-depth discussion of Wooden's coaching style and the lessons it may teach business leaders: Riggio, R.E. (2010), "The Leadership of John Wooden," *Psychology Today* blog, https://www.psychologytoday.com/blog/cutting-edge-leadership/201006/the-leadership-john-wooden.

31 Ames, N. (13 June 2016), "Meet Heimir Hallgrimsson, Iceland's Co-manager and Practicing Dentist," ESPN blog, http://www.espn.com/soccer/club/iceland/470/blog/post/2879337/meet-heimir-hallgrimsson-icelands-co-manager-and-practicing-dentist.

Chapter 10

1 See, for instance, Izon, D., Danenberger, E.P., and Mayes, M. (2007), "Absence of Fatalities in Blowouts Encouraging in MMS Study of OCS Incidents 1992–2006," *Drilling Contractor*, 63(4), 84–9. Gold, R. and Casselman, B. (30 April 2010), "Drilling Process Attracts Scrutiny in Rig Explosion," *Wall Street Journal*, 30.

2 https://www.theguardian.com/environment/2010/dec/07/transocean-oil-rig-north-sea-deepwater-horizon.

3 Vaughan, A. (16 January 2018), "BP's Deepwater Horizon Bill Tops $65bn," *Guardian*, https://www.theguardian.com/business/2018/jan/16/bps-deepwater-horizon-bill-tops-65bn.

4 Barstow, D., Rohde, D., and Saul, S. (25 December 2010), "Deepwater Horizon's Final Hours," *New York Times*, http://www.nytimes.com/2010/12/26/us/26spill

.html. And Goldenberg, S. (8 November 2010), "BP Had Little Defence against a Disaster, Federal Investigation Says," *Guardian*, https://www.theguardian.com/environment/2010/nov/08/bp-little-defence-deepwater-disaster.

5 Spicer, A. (2004), "Making a World View? Globalisation Discourse in a Public Broadcaster," PhD thesis, Department of Management, University of Melbourne, https://minerva-access.unimelb.edu.au/handle/11343/35838.

6 Alvesson, M., and Spicer, A. (2016), *The Stupidity Paradox: The Power and Pitfalls of Functional Stupidity at Work*, London: Profile, Kindle Edition (Kindle Locations 61–7).

7 Alvesson and Spicer, *The Stupidity Paradox* (Kindle Locations 192–8).

8 Grossman, Z. (2014), "Strategic Ignorance and the Robustness of Social Preferences," *Management Science*, 60(11), 2659–65.

9 Spicer has written about this research in more depth for *The Conversation*: Spicer, A. (2015), "'Fail Early, Fail Often' Mantra Forgets Entrepreneurs Fail to Learn," https://theconversation.com/fail-early-fail-often-mantra-forgets-entrepreneurs-fail-to-learn-51998.

10 Huy, Q., and Vuori, T. (2015), "Who Killed Nokia? Nokia Did," *INSEAD Knowledge*, https://knowledge.insead.edu/strategy/who-killed-nokia-nokia-did-4268. Vuori, T.O. and Huy, Q.N. (2016), "Distributed Attention and Shared Emotions in the Innovation Process: How Nokia Lost the Smartphone Battle," *Administrative Science Quarterly*, 61(1), 9–51.

11 Grossmann, I. (2017), "Wisdom in Context," *Perspectives on Psychological Science*, 12(2), 233–57; Staw, B.M., Sandelands, L.E., and Dutton, J.E. (1981), "Threat Rigidity Effects in Organizational Behavior: A Multilevel Analysis," *Administrative Science Quarterly*, 26(4), 501–24.

12 The entire slideshow is available here: https://www.slideshare.net/reed2001/culture-1798664.

13 Feynman, R. (1986), "Report of the Presidential Commission on the Space Shuttle Challenger Accident," Volume 2, Appendix F, https://spaceflight.nasa.gov/outreach/SignificantIncidents/assets/rogers_commission_report.pdf.

14 Dillon, R.L. and Tinsley, C.H. (2008), "How Near-misses Influence Decision Making under Risk: A Missed Opportunity for Learning," *Management Science*, 54(8), 1425–40.

15 Tinsley, C.H., Dillon, R.L., and Madsen, P.M. (2011), "How to Avoid Catastrophe," *Harvard Business Review*, 89(4), 90–7.

16 Accord, H., and Camry, T. (2013), "Near-Misses and Failure (Part 1)," *Harvard Business Review*, 89(4), 90–7.

17 Cole, R.E. (2011), "What Really Happened to Toyota?" *MIT Sloan Management Review*, 52(4).

18 Cole, "What Really Happened to Toyota?"

19 Dillon, R.L., Tinsley, C.H., Madsen, P.M., and Rogers, E.W. (2016), "Organizational Correctives for Improving Recognition of Near-Miss Events," *Journal of Management*, 42(3), 671–97. For more information, see the French Bureau of Enquiry and

Analysis for Civil Aviation Safety's official report on the accident, the English translation of which is available at https://www.bea.aero/docspa/2000/f-sc000725a/pdf/f-soc000725a.pdf.

20 Tinsley, Dillon, and Madsen, "How to Avoid Catastrophe."

21 Dillon, Tinsley, Madsen, and Rogers, "Organizational Correctives for Improving Recognition of Near-Miss Events."

22 Reader, T.W. and O'Connor, P. (2014), "The Deepwater Horizon Explosion: Non-Technical Skills, Safety Culture, and System Complexity," *Journal of Risk Research*, 17(3), 405–24. See also House of Representatives Committee on Energy and Commerce, Subcommittee on Oversight and Investigations (25 May 2010), "Memorandum 'Key Questions Arising From Inquiry into the Deepwater Horizon Gulf of Mexico Oil Spill,'" online at http://www.washingtonpost.com/wp-srv/photo/homepage/memo_bp_waxman.pdf.

23 Tinsley, C.H., Dillon, R.L., and Cronin, M.A. (2012), "How Near-Miss Events Amplify or Attenuate Risky Decision Making," *Management Science*, 58(9), 1596–1613.

24 Tinsley, Dillon, and Madsen, "How to Avoid Catastrophe."

25 National Commission on the BP Deepwater Horizon Oil Spill (2011), *Deep Water: The Gulf Oil Disaster and the Future of Offshore Drilling*, p. 224, https://www.gpo.gov/fdsys/pkg/GPO-OILCOMMISSION/pdf/GPO-OILCOMMISSION.pdf.

26 Deepwater Horizon Study Group (2011), "Final Report on the Investigation of the Macondo Well Blowout," Center for Catastrophic Risk Management, University of California at Berkeley, http://ccrm.berkeley.edu/pdfs_papers/bea_pdfs/dhsgfinalreport-march2011-tag.pdf.

27 Weick, K.E., Sutcliffe, K.M. and Obstfeld, D. (2008), "Organizing for High Reliability: Processes of Collective Mindfulness," *Crisis Management*, 3(1), 31–66. Plain-language explanations of the characteristics of mindful organizations were also inspired by the following paper: Sutcliffe, K.M. (2011), "High Reliability Organizations (HROs)," *Best Practice and Research Clinical Anaesthesiology*, 25(2), 133–44.

28 Bronstein, S., and Drash, W. (2010), "Rig Survivors: BP Ordered Shortcut on Day of Blast," CNN, http://edition.cnn.com/2010/US/06/08/oil.rig.warning.signs/index.html.

29 "The Loss of USS Thresher (SSN-593)" (2014), http://ussnautilus.org/blog/the-loss-of-uss-thresher-ssn-593/.

30 National Commission on the BP Deepwater Horizon Oil Spill (2011), *Deep Water: The Gulf Oil Disaster and the Future of Offshore Drilling*, p. 229, https://www.gpo.gov/fdsys/pkg/GPO-OILCOMMISSION/pdf/GPO-OILCOMMISSION.pdf.

31 Cochrane, B.S., Hagins Jr, M., Picciano, G., King, J.A., Marshall, D.A., Nelson, B., and Deao, C. (2017), "High Reliability in Healthcare: Creating the Culture and Mindset for Patient Safety," *Healthcare Management Forum*, 30(2), 61–8.

32 See the following paper for one example: Roberts, K.H., Madsen, P., Desai, V., and Van Stralen, D. (2005), "A Case of the Birth and Death of a High Reliability Healthcare Organisation," *BMJ Quality & Safety*, 14(3), 216–20. A further dis-

cussion can also be found here: Sutcliffe, K.M., Vogus, T.J., and Dane, E. (2016), "Mindfulness in Organizations: A Cross-Level Review," *Annual Review of Organizational Psychology and Organizational Behavior*, 3, 55–81.

33 Dweck, C. (2014), "Talent: How Companies Can Profit From a 'Growth Mindset'," *Harvard Business Review*, 92(11), 7.

34 National Commission on the BP Deepwater Horizon Oil Spill, *Deep Water*, p. 237.

35 Carter, J.P. (2006), "The Transformation of the Nuclear Power Industry," *IEEE Power and Energy Magazine*, 4(6), 25–33.

36 Koch, W. (20 April 2015), "Is Deepwater Drilling Safer, 5 Years after Worst Oil Spill?" *National Geographic*, https://news.nationalgeographic.com/2015/04/150420-bp-gulf-oil-spill-safety-five-years-later/. See also the following for a discussion of the oil industry's self-regulation, and the reasons it is not comparable to INPO: "An Update on Self-Regulation in the Oil Drilling Industry" (2012), *George Washington Journal of Energy and Environmental Law*, https://gwjeel.com/2012/02/08/an-update-on-self-regulation-in-the-oil-drilling-industry/.

37 Beyer, J., Trannum, H.C., Bakke, T., Hodson, P.V., and Collier, T.K. (2016), "Environmental Effects of the Deepwater Horizon Oil Spill: A Review," *Marine Pollution Bulletin*, 110(1), 28–51.

38 Lane, S.M., et al. (November 2015), "Reproductive Outcome and Survival of Common Bottlenose Dolphins Sampled in Barataria Bay, Louisiana, USA, Following the Deepwater Horizon Oil Spill," *Proceedings of the Royal Society B*, 282(1818), 20151944.

39 Jamail, D. (20 April 2012), "Gulf Seafood Deformities Alarm Scientists," Al Jazeera.com, https://www.aljazeera.com/indepth/features/2012/04/201241682318260912.html.

Epilogue

1 Besides the material covered in Chapters 1, 7 and 8, see the following: Jacobson, D., Parker, A., Spetzler, C., De Bruin, W.B., Hollenbeck, K., Heckerman, D., and Fischhoff, B. (2012), "Improved Learning in US History and Decision Competence with Decision-Focused Curriculum," *PLOS One*, 7(9), e45775.

2 Owens, B.P., Johnson, M.D., and Mitchell, T.R. (2013), "Expressed Humility in Organizations: Implications for Performance, Teams, and Leadership," *Organization Science*, 24(5), 1517–38.

3 Sternberg, R.J. (in press), "Race to Samarra: The Critical Importance of Wisdom in the World Today," in Sternberg, R.J. and Glueck, J. (eds), *Cambridge Handbook of Wisdom* (2nd edn), New York: Cambridge University Press.

4 Howell, L. (2013), "Digital Wildfires in a Hyperconnected World," *WEF Report*, 3, 15–94.

5 Wang, H., and Li, J. (2015), "How Trait Curiosity Influences Psychological Well-Being and Emotional Exhaustion: The Mediating Role of Personal Initiative," *Personality and Individual Differences*, 75, 135–40.

Acknowledgments

This book could not exist if it were not for the generosity of many people. Thanks most of all to my agent, Carrie Plitt, for her enthusiastic belief in my proposal and her support and guidance ever since. Thanks also to the rest of the team at Felicity Bryan Associates, Zoë Pagnamenta in New York, and the team at Andrew Nurnberg Associates for helping to spread the word in the rest of the world.

I have been lucky to have been guided by my editors, Drummond Moir at Hodder & Stoughton and Matt Weiland at W. W. Norton. Their wise judgment and tactful edits have improved this book no end and I have learned so much from their advice. Thanks also to Cameron Myers at Hodder for his suggestions and for all his help in ensuring the editorial process flowed as smoothly as possible.

I am incredibly grateful to the many experts who shared their insights and knowledge with me. These include: David Perkins, Robert Sternberg, James Flynn, Keith Stanovich, Wändi Bruine de Bruin, Dan Kahan, Hugo Mercier, Itiel Dror, Rohan Williamson, Igor Grossmann, Ethan Kross, Andrew Hafenbrack, Silvia Mamede, Pat Croskerry, Norbert Schwarz, Eryn Newman, Gordon Pennycook, Michael Shermer, Stephan Lewandowsky, John Cook, Susan Engel, Carol Dweck, Tenelle Porter, James Stigler, Robert and Elizabeth Bjork, Ellen Langer, Anita Williams Woolley, Angus Hildreth, Bradley Owens, Amy Yi Ou, Andre Spicer, Catherine Tinsley, Karlene Roberts—and the many other interviewees who have not been quoted, but whose expertise nevertheless contributed to my argument.

Thanks also to Brandon Mayfield for kindly sharing his experiences with me; Michael Story, who gave me a little glimpse of what it means to be a super-forecaster; and Jonny Davidson for his help with the diagrams. I am also grateful to the staff and students at the Intellectual Virtues Academy in Long Beach, who could not have been more welcoming during my visit.

Richard Fisher at BBC Future first commissioned me to write a piece on the "downsides of being intelligent" in 2015. Thanks for setting the ball in motion, and for the continued encouragement and advice throughout my career. And thanks to my friends and colleagues, including Sally Adee, Eileen and Peter Davies, Kate Douglas, Stephen Dowling, Natasha and Sam Fenwick, Simon Frantz, Melissa Hogenboom, Olivia Howitt, Christian Jarrett, Emma and Sam Partington, Jo Perry, Alex Riley, Matthew Robson, Neil and Lauren Sullivan, Helen Thomson, Richard Webb, and Clare Wilson, who have all offered invaluable support. I owe you all many drinks. *A Marta, Luca, Damiano e Stefania, grazie infinite.*

I owe more than I can describe to my parents, Margaret and Albert, and to Robert Davies, for your support in every step of this journey. I could not have written this book without you.

Illustration Credits

The publisher would like to thank the following for their permission to reproduce these images. Every attempt has been made to contact the copyright holders, however if any omissions have occurred please contact the publishers and we will rectify any errors at the earliest opportunity.

Page 15: Courtesy of National Library of Medicine/NCBI

Page 19: Wikimedia/Life of Riley. CC BY-SA 3.0 https://creativecommons.org/licenses/by-sa/3.0/deed.en

Page 25: Our World in Data/Max Roser CC BY-SA https://creativecommons.org/licenses/by-sa/3.0/au/

Page 55: Kahan, D.M. "Ordinary science intelligence": a science-comprehension measure for study of risk and science communication, with notes on evolution and climate change, J. Risk Ress 20, 995–1016 (2017).

Page 78: Courtesy of the US Department of Justice's Office of the Inspector General's report "Review of the FBI's Handling of the Brandon Mayfield Case" https://oig.justice.gov/special/s1105.pdf

Page 136: Diagram and captions created by the author

Pages 227–8: Roderick I. Swaab, Michael Schaerer, Eric M. Anicich, et al. "The Too-Much-Talent Effect: Team Interdependence Determines When More Talent Is Too Much or Not Enough" *Psychological Science* 25(8), 1581–1591 (2014). Reprinted by Permission of SAGE Publications, Inc.

Index